CW00448079

OXFORD CHEMISTRY PRIMERS

Physical Chemistry Editor	Founding/Organic Editor	Inorganic Chemistry Editor	Chemical Engineering Editor
RICHARD G. COMPTON	STEPHEN G. DAVIES	JOHN EVANS	LYNN F. GLADDEN
University of Oxford	University of Oxford	University of Southampton	University of Cambridge

Structure and Reactivity in Organic Chemistry

Howard Maskill

Senior Lecturer in Chemistry at
the University of Newcastle upon Tyne

OXFORD

UNIVERSITY PRESS

This book has been printed digitally and produced in a standard specification
in order to ensure its continuing availability

OXFORD
UNIVERSITY PRESS

Great Clarendon Street, Oxford OX2 6DP

Oxford University Press is a department of the University of Oxford.
It furthers the University's objective of excellence in research, scholarship,
and education by publishing worldwide in

Oxford New York

Auckland   Cape Town Dar es Salaam Hong Kong Karachi
Kuala Lumpur Madrid Melbourne Mexico City Nairobi
New Delhi Shanghai Taipei Toronto
With offices in
Argentina Austria Brazil Chile Czech Republic France Greece
Guatemala Hungary Italy Japan South Korea Poland Portugal
Singapore Switzerland Thailand Turkey Ukraine Vietnam

Oxford is a registered trade mark of Oxford University Press
in the UK and in certain other countries

Published in the United States
by Inc., New York

© Howard Maskill 1999

Not to be reprinted without permission
The moral rights of the author have been asserted
Database right Oxford University Press (maker)

Reprinted 2005

All rights reserved. No part of this publication may be reproduced,
stored in a retrieval system, or transmitted, in any form or by any means,
without the prior permission in writing of Oxford University Press,
or as expressly permitted by law, or under terms agreed with the appropriate
reprographics rights organization. Enquiries concerning reproduction
outside the scope of the above should be sent to the Rights Department,
Oxford University Press, at the address above

You must not circulate this book in any other binding or cover
And you must impose this same condition on any acquirer

ISBN 978-0-19-855820-0

Series Editor's Foreword

The foundations of our mechanistic understanding of organic chemistry are based on interpretation of how molecular structure affects chemical reactivity. The kinetics and thermodynamics from which mechanistic information is gleaned include experimental data from studies of equilibria, reaction rates, correlation analysis, catalysis and isotope effects. Howard Maskill covers all these topics in this Primer which is the companion to his Mechanisms of Organic Reactions (Primer No 45).

Oxford Chemistry Primers have been designed to provide concise introductions to all students of chemistry and contain only the essential material that would normally be included in an 8–10 lecture course. The excellent pedagogical description of physical organic chemistry in this Primer will be of interest to apprentice and master chemists alike.

Stephen G. Davies
Dyson Perrins Laboratory,
University of Oxford

Preface

The nature and structure of an organic molecule are described in terms of its valence electrons (σ bonds, π bonds, and lone pairs), bond lengths, bond angles, and bond polarities. Chemical reactivity is expressed by rate constants and, for reversible reactions, equilibrium constants under specified experimental conditions. The relationships between the nature and structure of a molecule and the compound's chemical reactivity, and how these relationships are investigated, are the subject of this book. It is an area of organic chemistry which requires a sound appreciation of some physical chemistry, and is seldom covered in general organic chemistry texts. My aim has been to help the reader to progress from the curly arrow approach to organic reaction mechanisms (a basic knowledge of which is assumed) towards quantitative considerations of reactivity based upon knowledge of molecular properties.

The first chapter introduces a way of describing organic chemical reactions which allows an analysis of the relationship between molecular structure and reactivity. Thereafter, I cover the principal methodologies by which reactivity relationships are investigated, and have included a broad range of results and applications. Throughout, I have tried to remain close to the experimental basis of the subjects covered and to avoid the desiccating effects of over-mathematical formalisms. Problems are provided at the ends of chapters.

All chapters have been read by experts on particular topics (Rory More O'Ferrall, Mike Page, John Shorter, Andrew Williams, and Ian Williams), and the whole text has been read by Ian Watt and Steve Davies. I am very grateful to these friends and colleagues for their help and guidance. I also wish to thank participants (students and teachers) on the Winter Schools on Organic Reactivity which have taken place since 1991 in Bressanone, Italy, under the direction of Gianfranco Scorrano. Much of this little book has been influenced by lectures and discussions which took place throughout these Winter Schools. Next, I must thank those chemists (past and present) whose experimental results I have used throughout the text in various ways. Strict limitations on space have prevented me from referencing their work in the usual manner. Finally, I would be grateful to hear from readers who detect errors of any sort in the book.

Newcastle upon Tyne   H. M.
March 1999

In loving memory of my parents, Rita and Walter Maskill.

Contents

1 Organic reaction mechanisms and reaction maps

1.1 Introduction

Collisions between molecules provide the energy for bimolecular and unimolecular thermal reactions to occur. For two molecules in the gas phase, there are six translational degrees of freedom and up to six rotational ones (depending upon the symmetries of the molecules). The number of vibrational degrees of freedom is then determined by the number of atoms in the two molecules such that the total number of degrees of freedom (translational plus rotational plus vibrational) is equal to $3n$ where n is the total number of atoms in the two molecules.

In a bimolecular reaction, three translational and (normally) up to three rotational degrees of freedom are lost in the formation of the activated complex; since the total has to remain the same, a corresponding number of new vibrational degrees of freedom are formed. In other words, translational and possibly rotational energy of the two molecules immediately before the collision is converted into vibrational energy in the rebonding which takes place as the original molecules become the activated complex. If insufficient energy of the colliding molecules is converted, the molecules simply distort then separate without reacting. Even if sufficient translational and rotational energy is converted into vibrational energy, the activated complex may still fall apart to reactant molecules rather than proceed to product(s).

In a unimolecular reaction, a molecular collision leaves one of the molecules in a vibrationally excited state which either reacts or simply loses its excess energy in a subsequent collision. In such a reaction, the molecular collision which produces the vibrationally excited reactant molecule may be between two molecules of the reactant, or between one of the reactant and a molecule of a different compound which does not (or cannot) react. Clearly, therefore, both unimolecular and bimolecular mechanisms involve the conversion of translational (and possibly rotational) energy into vibrational energy. Consequently, an understanding of intermolecular interactions and vibrational properties of molecules is essential for an understanding of organic reaction mechanisms.

In photochemical reactions, molecules are energised by the absorption of electromagnetic radiation, and electronically excited states are involved.

The *total* number of degrees of freedom of a system of atoms depends only upon the number of atoms, *n*. How the $3n$ degrees of freedom are distributed amongst translation, vibration, and rotation depends upon how the atoms are combined into molecules.

The term *activated complex* is discussed in the next chapter.

$$A + B \underset{(\leftarrow)}{\longrightarrow} AB^{\ddagger} \longrightarrow Product(s)$$

Bimolecular mechanism

$$A + B \rightleftharpoons A^* + B$$
$$A^* \longrightarrow Product(s)$$

Unimolecular mechanism

The Lindemann-Hinshelwood mechanism accounts for how gas phase unimolecular reactions in which the activated molecule is produced in a bimolecular collision may lead to either a first- or a second-order rate law.

1.2 Molecular vibrations and potential energy diagrams

Two molecular potential energy curves for a diatomic molecule A–B are shown in Fig. 1.1; one incorporates the simple harmonic approximation with two quantised energy levels included, and the other is anharmonic. They are alternative descriptions of the relationship between the potential energy of the diatomic molecule and a single structural variable which can be identified precisely–the internuclear distance, r. The same relationships can be described mathematically.

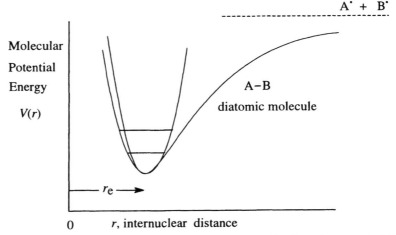

For the simple harmonic approximation,
$$V(r) = 0.5\kappa(r - r_e)^2$$
and
$$\kappa = \frac{d^2V}{dr^2}$$

Fig. 1.1 Harmonic and anharmonic potential energy curves for diatomic molecule A-B

The symmetrical simple harmonic curve is characterised by just two parameters, the force constant, κ, which is a measure of the strength of the bond and is reflected by the narrowness of the curve, i.e. the curvature, and the equilibrium bond length, r_e. Whilst this is a helpful starting point, it is inadequate as a model for the behaviour of real molecules. It gives increasingly poor descriptions of molecular vibrations as the vibrational amplitude increases, and cannot account for the dissociation of the molecule.

The unsymmetrical curve in Fig. 1.1 shows that both compression and stretching of the bond lead to anharmonicity, and the force constant (curvature) now depends upon the bond length. A third parameter has to be added to the mathematical relationship between structure and energy to account for the anharmonicity. At ever smaller displacements from the equilibrium bond-length, however, the difference between the harmonic and anharmonic curves becomes vanishingly small.

Figure 1.1 describes the homolysis of a diatomic molecule A-B in the gas phase (or, equivalently, the bonding of the two atoms A and B as they approach and form the diatomic molecule), and an analogous curve may be constructed to describe heterolysis and its reverse.

The qualitative use of such diagrams may also be extended to describe the stretching (and possible cleavage) of bonds in polyatomic molecules. However, the structural variable is now a complex function which is usually approximated by its main component. Figure 1.2 comprises sketches describing the homolysis and heterolysis of chloromethane. In both, the methyl is treated as a single group. However, it has internal structure and the internal equilibrium configuration will change as the C–Cl bond stretches and ultimately cleaves. Consequently, we refer to a composite *reaction coordinate, q,* rather than simply the C·····Cl internuclear distance. And the internal changes of the methyl group which necessarily accompany the stretching of the carbon–chlorine bond will be different according to whether the cleavage is homolytic or heterolytic.

The energy cost of separating opposite electrical charges in the gas phase is higher than that of separating a pair of uncharged groups, so the asymptote corresponding to dissociated ions is at higher energy than that for dissociated atoms or radicals.

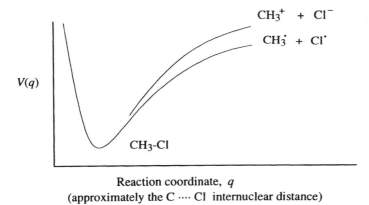

Fig. 1.2 Potential energy diagram for the gas phase homolysis and heterolysis of CH_3Cl

Potential energy curves for dissociations involving multibond processes, even if they are concerted, are qualitatively different as we shall see later.

1.3 Noncovalent intermolecular interactions and bimolecular encounter complexes

The potential energy curves of Figs 1.1 and 1.2 represent the formation of covalent bonds between atoms or groups as well as the dissociation of stable molecules into fragments, and the energies involved are large; covalent bond energies are typically hundreds of kilojoules per mole. However, there are interactions between atoms, ions, and molecules which do not lead to the formation of covalent bonds and which individually are relatively weak. The most familiar is the hydrogen bond which forms between a weakly acidic hydrogen, e.g. of an alcohol or water, and the lone pair of a Lewis base such as an amine, or another water or alcohol molecule. There are other types of weak interactions such as between an ion and a dipolar molecule, and between

The enthalpy of a hydrogen bond between an OH and a lone pair on another oxygen is between about 10 and 30 kJ mol⁻¹. This compares with 381 kJ mol⁻¹ for the C-O bond in methanol and 427 for the O–H bond.

two dipolar molecules. These ion-dipole and dipole-dipole interactions are, like hydrogen bonds, essentially electrostatic in origin. There are also even weaker interactions between nonpolar molecules which arise from the polarisability of the valence electrons of the molecules rather than permanent charges or dipoles. All these weak intermolecular interactions are zero at large intermolecular separations, attractive (cohesive) at distances comparable with typical bond lengths, then increasingly strongly repulsive at increasingly close intermolecular distances. A representative intermolecular interaction energy curve is sketched in Fig. 1.3 for two molecules Y and Z.

The interaction between Y and Z at large intermolecular separations is zero, then favourable interactions lead to the formation of a bimolecular encounter complex, Y·Z. Forcing Y and Z closer together leads to increasingly strong repulsion. The strength of the bonding at the minimum in the profile depends upon the nature of the interaction, and the particular example. The minimum could be very shallow representing a weakly bonded complex of very short lifetime.

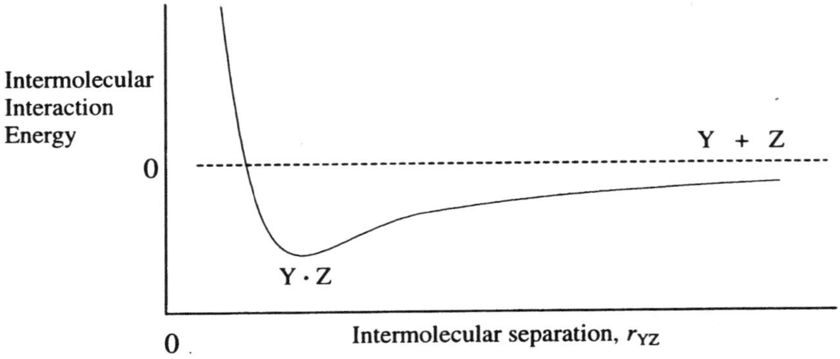

Fig. 1.3 Energy profile for a weak noncovalent intermolecular interaction

Whilst all such interactions singly are very weak compared with typical covalent bonds, they are additive so two large molecules could bind strongly through the cooperative effects of many noncovalent interactions. This is often the mechanism by which biomolecules bind, e.g. the two strands of DNA or, less strongly, an enzyme with its substrate prior to reaction, and chemical messengers with their receptor sites in physiological processes.

In the present context, noncovalent interactions are important as they are involved in the early stages of bimolecular reactions, i.e. in the initial translational stage before the redistribution of valence electrons occurs.

1.4 Potential energy surfaces for intermolecular group transfer with a single vibrational reaction coordinate

A very common type of chemical reaction is the transfer of a small labile group X from one larger residue A to another M, eqn 1.1.

X could be hydrogen or methyl; A and M are either much larger atoms or polyatomic groups which may be treated for the present as though they are single large atoms.

$$A-X \ + \ M \ \rightleftharpoons \ A \ + \ X-M \qquad (1.1)$$

An outline of one possible mechanism for this transfer is shown in eqn 1.2.

$$A\text{-}X \ + \ M \ \rightleftharpoons \ A\text{-}X\cdot M \ \rightleftharpoons \ A\cdot X\text{-}M \ \rightleftharpoons \ A \ + \ X\text{-}M \qquad (1.2)$$

First, we have the association of AX and M to give an encounter complex, then the transfer of X from A to M within the encounter complex to give an isomeric complex, then finally this complex dissociates. The first and final steps correspond to translations whereas the transfer of X from A to M within a bimolecular encounter complex may be seen as a vibration, and a potential energy profile for it can be constructed from the separate anharmonic potential energy curves for A-X and M-X. The individual profiles will be similar to that in Fig. 1.1. However, the proximity of M to AX in the encounter complex will perturb the vibration of AX and, correspondingly, the proximity of A to MX following the transfer of X will perturb the vibration of MX. In Fig. 1.4, the two potential energy curves, each modified to account for these intermolecular interactions, are combined to describe the transfer of X from A to M within the encounter complex.

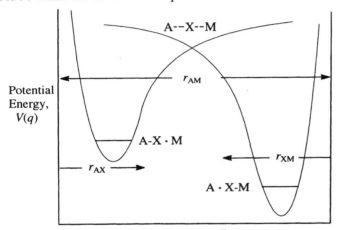

Figure 1.4 and the associated text refer to a transfer of X between A and M with the three groups collinear. This ensures that only a single structural variable (the reaction coordinate) is required to define the location of X between A and M. Thus, the reaction coordinate of Fig. 1.4 is the location of X between A and M.

Fig. 1.4 Combined potential energy curves for A-X and X-M providing a model for the collinear transfer of X between A and M within an encounter complex

We see in Fig. 1.4 that the point at which the two curves cross corresponds to a configuration A····X····M in which X is partly and equally bonded to A and to M. From this point, X moves towards A or M and the potential energy of the system decreases; the upwardly-continuing curves to the left and right from the intersection have no significance in the combined diagram which is redrawn in Fig. 1.5. In this, we have included three more parabolas in an additional dimension (in and out of the plane of the page) to acknowledge that A, X, and M may not all be single atoms, but groups with internal structure.

The parabola superimposed upon the curve for the vibration of AX is the molecular potential energy curve for a representative vibration which is not a significant component of the reaction coordinate, i.e. an internal vibration of A, X, or M. This vibration is transformed as the group transfer takes place, and is represented by the shallower parabola for the transition structure. It is transformed further as the group transfer becomes complete, and represented by the parabola superimposed upon the potential energy curve for the vibration of XM. The continuous curve which begins by describing a stretching vibration for AX, then passes over a maximum corresponding to

The equal bonding of X to A and to M does not imply that the distances A·····X and X·····M are equal.

the transition structure, and ends up describing the stretching vibration of XM, is the reaction coordinate for the group transfer.

The reaction coordinate in Fig. 1.5, as in Fig. 1.4, is the location of X between A and M.

Fig. 1.5 Molecular potential energy profile for transfer of X between A and M within an encounter complex with superimposed parabolas corresponding to an internal vibration

We have included the three parabolas only at the minima and maximum in the reaction coordinate, but they represent sections through a potential energy surface with hollows corresponding to the reactant and product complexes, and a saddle point corresponding to the transition structure. Dimensional restrictions prevent more than just one internal vibration being represented in diagrams such as Fig. 1.5. Depending upon the molecular complexity of A, X, and M, however, there may be many more. We can now only imagine the multi-dimensional potential energy hypersurface for this group transfer within the encounter complex. It is characterised by multi-dimensional mimima corresponding to reactant and product. The hypersurface also includes a single saddle point corresponding to the transition structure for which all internal degrees of freedom except one have positive force constants, and one (corresponding to the unique degree of freedom which is the reaction coordinate) has a negative force constant.

Potential energy curves

$\kappa > 0$ $\kappa < 0$

A negative force constant, by eqn 5.6 on p. 84, corresponds to an imaginary vibrational frequency.

So far, we have considered qualitatively, but in detail, the changes which occur within the encounter complex leading to a group transfer; the reaction coordinate is vibrational. We also saw in the previous section that the formation of an encounter complex may be described by a different type of profile in which the reaction coordinate is translational. We can now combine these and construct in Fig. 1.6 an overall reaction profile for the reaction of eqn 1.1 by the mechanism outlined in eqn 1.2. In Fig. 1.6, we are obliged to acknowledge that the nature of the reaction coordinate changes as the reaction progresses.

This description of the reaction of eqn 1.1 has allowed a potential energy profile for the group transfer to be constructed which includes a single though composite reaction coordinate. The second configurational coordinate available to us could be used to represent a vibration which is not part of the reaction coordinate as in Fig. 1.5. This model will prove useful when we consider deuterium kinetic isotope effects in chapter 5.

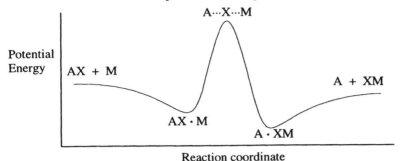

Fig. 1.6 Composite reaction profile for the reaction of eqn 1.1 by the mechanism of 1.2

Note how the reaction coordinate in Fig. 1.6 is initially a translational degree of freedom, then vibrational, and finally translational again as the reaction proceeds to completion.

More often, however, we shall wish to analyse effects of structural modifications to A, X, and M upon the reaction and, for this, use of more than a single composite reaction coordinate is desirable. We shall wish to consider two independent components of the overall translational/vibrational reaction coordinate, and that requires an alternative analysis of the same overall transformation.

1.5 Potential energy surfaces for intermolecular group transfer with dissected vibrational reaction coordinates

We shall now construct a three-dimensional diagram for the reaction of eqn 1.1 which allows us to describe changes in the potential energy of the A, X, M system as A····X and X····M distances vary independently.

$$A-X \; + \; M \; \rightleftharpoons \; A + X-M \qquad (1.1)$$

In Fig. 1.7, the vertical axis in the plane of the page (the y-axis) is the potential energy of the collinear A, X, M system, the horizontal axis in the plane of the page (the x-axis) is the distance from X to M, and the horizontal axis extending behind the plane of the page (the z-axis) is the distance from X to A.

The plane defined by the points y, x, and z in Fig. 1.7 includes structures with the same M····X distance. The relationship between structure and potential energy of all A–X–M configurations within this y-x-z plane, where the M····X distance is greater than typical bond lengths, is the anharmonic potential energy profile for the A-X molecule perturbed by M at a fixed distance from X shown in Fig. 1.8. Moving this plane to the left or right in Fig. 1.7 alters the M····X distance and this has an effect upon the A-X potential energy curve. One can see that a large number of potential energy curves like that in Fig. 1.8 corresponding to a large number of y-x-z planes in Fig. 1.7 describe part of a three-dimensional potential energy surface of the A, X, M system.

The system is restricted to a collinear arrangement of A, X, and M to ensure that a configuration defined by two coordinates (an A····X distance and an X····M distance) is unique.

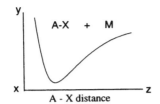

Fig. 1.8 Potential energy profile for the A–X stretching vibration at a constant X$\cdots$M distance (greater than an X–M bond length), i.e. a section through the energy surface of Fig. 1.7 in the y-x-z plane

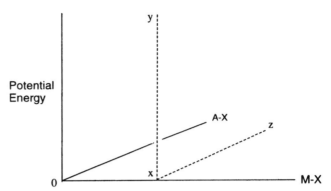

Fig. 1.7 Coordinates for a three-dimensional potential energy diagram for the collinear A, X, M system including the y-x-z plane defining a constant M–X distance

In both Figs 1.7 and 1.9, the configurational origin corresponds to the superposition of A, X, and M the energy of which is infinite.

The same axes are drawn in Fig. 1.9 but a different plane is now defined by points y', z', and x', and all configurations within this plane have the same A$\cdots$X distance (greater than the equilibrium A–X bond length). The only structural variable for configurations within this plane, therefore, is the M$\cdots$X distance, and the stretching vibration potential energy curve for the M-X molecule at the constant distance from A to X is shown in Fig. 1.10.

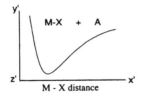

Fig. 1.10 Potential energy profile for the M–X stretching vibration at a constant X$\cdots$A distance (greater than an X–A bond length), i.e. a section through the energy surface of Fig. 1.9 in the y'-z'-x' plane

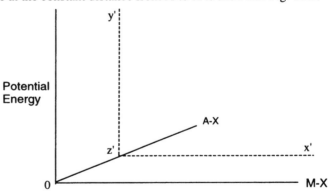

Fig. 1.9 Coordinates for the three-dimensional potential energy diagram for the collinear A, X, M system including the y'-z'-x' plane defining a constant A–X distance

We can now see that a large number of potential energy curves like that in Fig. 1.10 corresponding to a large number of y'-z'-x' planes in Fig. 1.9 describe another part of the three-dimensional potential energy surface of the A, X, M system. By combining our thoughts on Figs 1.7 and 1.9, we can visualise the complete three-dimensional potential energy surface for all collinear configurations of A, X, and M within this coordinate system.

A plane x"-y"-z" at constant high energy and another x'''-y'''-z''' at constant low energy are shown in Fig. 1.11. These cut through the potential energy surface of the A, X, M system as indicated.

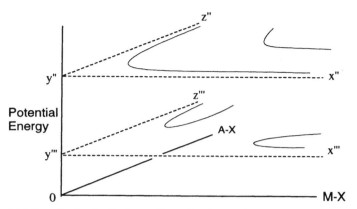

Fig. 1.11 Coordinates for the three-dimensional potential energy diagram for the collinear A, X, M system including two planes of constant energy

The shapes and details of the potential energy curves in Figs 1.8 and 1.10 will depend upon the proximity of M to AX and of A to MX, i.e. how far out from the origin the y-x-z and y'-z'-x' planes are drawn. The potential energy curve contained by a plane y-x-z far from the origin corresponds to the potential energy curve of molecule A-X unaffected by M. Correspondingly, the potential energy curve contained by a plane y'-z'-x' far from the origin corresponds to the potential energy curve of molecule M-X unaffected by A.

1.6 Reaction maps

By projecting the lines in Fig. 1.11 which join together points corresponding to configurations of equal energy onto the base of Fig. 1.11, and adding others, we generate Fig. 1.12; this is a two-dimensional potential energy contour map of the collinear A, X, M system. Superimposed on this energy contour diagram are various features which require explanation.

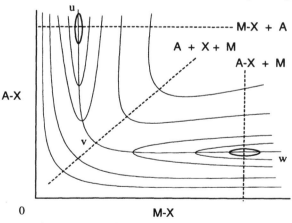

Fig. 1.12 Potential energy contour map for the group transfer of eqn 1.1

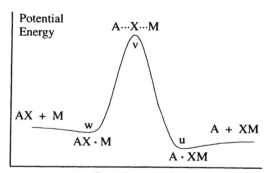

Fig. 1.13 A section through the surface of Fig. 1.12 along the curved trough through u, v, and w. As expected, Fig. 1.13 is very similar to Fig. 1.6 as they both describe the mechanism of eqn 1.2 even though they have been constructed differently.

The dotted line along the top of Fig. 1.12 represents a section through the energy surface corresponding to a vibration of the M-X molecule with A at a constant distance from X and is described by the potential energy curve in Fig. 1.10. Analogously, the dotted line up the right hand side of Fig. 1.12 represents a section through the surface corresponding to the vibration of A-X with M at a constant distance from X, and is described by the potential energy curve shown in Fig. 1.8. The curved line u-v-w is drawn along the trough of the potential energy surface and is the minimum energy path between the configurations represented by u and w, i.e. the reaction of eqn 1.1. Point v on this trajectory is a saddle point on the surface and corresponds to the transition structure for the group transfer. The dotted diagonal line extending from the origin corresponds to configurations in which the A····X distance is equal to the X····M distance. At one very high energy extreme, it leads to the superposition of the groups A, X, and M at the origin; at the other end, it extends towards very large and equal separations of the three groups. There is a minimum in this section through the surface at v, and small displacements within this section in either direction at v correspond to the symmetrical stretching vibration of the transition structure, a vibration orthogonal to the reaction coordinate at v.

Figure 1.14(a) is a rectangle cut out of the middle of Fig. 1.12 with the top left corresponding to the molecule M-X close to A, and the bottom right corresponds to molecule A-X close to M. Figure 1.14(b) is the same rectangle reflected from left to right. The bottom left corner of the rectangle now corresponds to A-X + M and the top right corresponds to A + X-M; i.e. from bottom left corner to top right of this configurational map corresponds to the chemical reaction of eqn 1.1. Moreover, the contours indicate a mechanism. M follows the lowest energy route corresponding to progression along the bottom of the map and approaches X to give a transition structure A···X···M; A departs by the lowest energy route up the right hand side of the diagram. This transformation involves a single transition structure, so is by definition a *concerted* process, but the development of bonding from M to X runs ahead of the unbonding of A from X. It is, therefore, an *associative asynchronous* concerted mechanism. We are now in a position to redraw Fig. 1.14(b) to consider other possible routes between the bottom left and the top right, i.e. alternative mechanisms for the reaction of eqn 1.1.

From bottom left to top right is the conventional way to use reaction maps to describe chemical equations written from left to right.

To keep the maps simple, contours are not labelled, but it should be understood that reactants and products are minima and all paths from them are initially up-hill.

$$A-X \; + \; M \; \rightleftharpoons \; A \; + \; X-M \qquad (1.12)$$

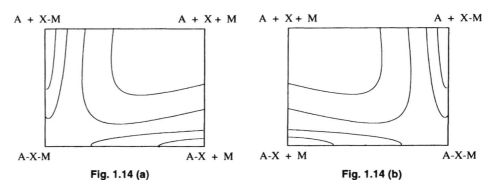

Fig. 1.14 (a)   Fig. 1.14 (b)

Reaction maps describing the reaction of eqn 1.1 by an associative asynchronous concerted mechanism

In Fig. 1.15(a), the lowest energy route across the map corresponds to progression up the left hand side, i.e. unbonding of A from X, followed by bonding of X to M, and the open transition structure is indicated by ‡. This remains a concerted mechanism (single transition structure) and is also asynchronous, but it is now *dissociative* as unbonding precedes bonding.

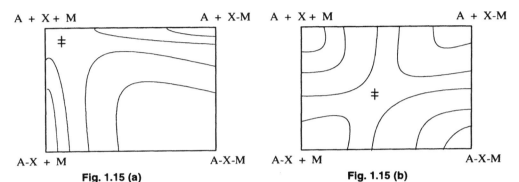

Fig. 1.15 (a)   **Fig. 1.15 (b)**

Reaction maps for concerted mechanisms of eqn 1.1 which are (a) dissociative asynchronous and (b) synchronous

The contours in Fig. 1.15(b) describe another mechanism in which the lowest energy route is across the centre of the reaction map and corresponds to *synchronous* concerted bonding of M to X and unbonding of A.

1.7 Parallel and perpendicular effects

Parallel effects. The contours in Fig. 1.15(b) indicate that the reaction is mildly exothermic and the transition structure is just less than half way along the reaction path. Making the reaction more exothermic, e.g. by some structural modification to one of the reactants, corresponds to a relative lowering of the energy of the top right of Fig. 1.15(b). Figure 1.16 includes a section through the energy surface along the reaction coordinate. We see that lowering the energy of the products with respect to reactants (e.g. by modifying M to give M') moves the position of the transition structure within the reaction coordinate towards the reactants. Thus, widening the energy gap moves the transition structure towards the higher energy end. Conversely, reducing the energy gap between reactants and products moves the transition structure towards the lower energy end. These effects of changing the exothermicity of the reaction upon the position of the transition structure *within the reaction coordinate* are known as *parallel* (or Hammond) effects.

Perpendicular effects. A section through the energy surface described by Fig. 1.15(b) from the top left, through the saddle point corresponding to the transition structure, to the bottom right is U-shaped, Fig. 1.17. The contours indicate that the energy of the bottom right is about the same as that of the top left. The section passes through the transition structure (which is about mid

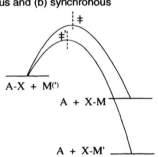

Fig. 1.16 Parallel effect of the exothermicity of the reaction in Fig. 1.15(b) upon the position of the transition structure within the reaction coordinate

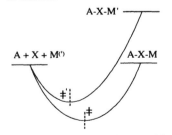

Fig. 1.17 Movement of the saddle point (transition structure) perpendicular to the reaction coordinate in Fig. 1.15(b)

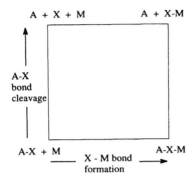

Fig. 1.18 Reaction map template for the reaction

$$A\text{-}X + M \rightleftharpoons A + X\text{-}M$$

way between the top left and the bottom right) perpendicular to the reaction coordinate. If a structural change is made to one of the reactants (M becomes M') which relatively raises the bottom right, we see that the saddle point moves towards the lower energy end (top left). Any further change which reduces the gap between top left and bottom right would move the saddle point back towards the higher end. In both cases, the movement of the transition structure is *perpendicular to the reaction coordinate* at the saddle point hence these are called *perpendicular* (or Thornton) effects.

1.8 Step-wise reactions

The hypothetical reactions so far considered have all been single-step concerted processes and have differed only in the degree of synchrony (the synchroneity) amongst the various changes. Structural modifications which reduce the energy of the top left state in Fig. 1.15(a) could actually generate a local minimum in that part of the potential energy surface. A generic example is a substitution reaction when the alkyl group becomes capable of forming a viable carbenium ion, R^+, as in eqn 1.3.

$$Y^- + R\text{-}X = [Y^-\cdot R^+\cdot X^-] = R\text{-}Y + X^- \qquad (1.3)$$

In this event, the reaction map is as in Fig. 1.19(a) and the intermediate is flanked in the reaction coordinate by two barriers, i.e. transition structures ‡1 and ‡2 now separate the intermediate from reactants and from products.

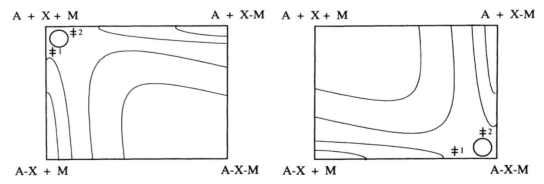

Fig. 1.19(a) Step-wise dissociative substitution mechanism **Fig. 1.19(b)** Step-wise addition-elimination substitution mechanism

Sections through both Fig. 1.19 (a) and (b) along the curved reaction paths lead to the following profile with ‡1 or ‡2 rate limiting according to the particular reaction.

Correspondingly, Fig. 1.19(b) describes a two-step addition-elimination mechanism, e.g. an acyl transfer via a tetrahedral intermediate, eqn 1.4.

1.9 Single-step multibond cycloaddition reactions

The cycloaddition reaction between buta-1,3-diene and acrolein (propenal) is shown in the dissociation direction in eqn 1.5.

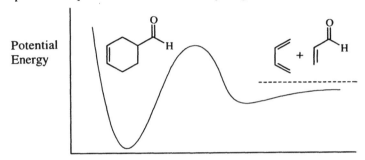

$$(1.5)$$

The molecular potential energy profile for this reaction is shown in Fig. 1.20 and should be contrasted with the ones in Figs 1.1 and 1.2 at the beginning of the chapter for simple reactions in which only single bonds cleave.

Potential
Energy

Reaction Coordinate

Fig. 1.20 Reaction profile for the retro-cycloaddition reaction of eqn 1.5

The reaction coordinate in Fig. 1.20 is vibrational on the left hand side and translational on the right.

The cycloaddition involves initial intermolecular association with a small (favourable) decrease in energy to form a bimolecular encounter complex; the reaction coordinate is translational. There is then an appreciable barrier corresponding to rehybridisation of four carbon atoms and changes in the relative positions of most atoms within the encounter complex as redistribution of the valence electrons takes place. During this process, the reaction coordinate is transformed into a vibrational degree of freedom.

In the dissociation direction, the reaction coordinate is initially a vibration of the adduct which increases in amplitude when the molecule gains energy through intermolecular collisions. At the barrier maximum, the vibration is becoming a translation then the energy decreases with the extensive reorganisation of valence electrons and formation of two stable molecules.

For other cycloadditions, alternative step-wise paths via reactive intermediates can be included in the reaction maps.

A reaction map template can be drawn with formation of the new σ-bonds as the two configurational coordinates. Synchronous and asynchronous paths can be included just as for the group transfer reactions considered earlier.

1.10 Conversion from the molecular to the molar scale

So far, our considerations have been of molecules, molecular potential energy, and molecular potential energy profiles and maps. The potential energy profile or hypersurface relates to individual isolated molecules at zero kelvin. However, when we investigate reaction mechanisms, we measure rate constants, equilibrium constants, and other parameters of reactions on macroscopic amounts. There are several stages involved in relating our theoretical considerations of individual molecules to the molar scale of real experiments. First, we need to scale up by the Avogadro number. Secondly, we have to acknowledge that, depending upon the temperature, higher vibrational and rotational levels will be populated, and molecules will have translational energy. Thirdly, we need to decide which standard state is most convenient for our macroscopic scale, e.g. if the reaction is in the gas phase, is it to be at 1 mol dm^{-3} or 1 atm? Fourthly, we need to take account of entropy aspects which become dominant at higher temperatures. Finally, and most difficult at the present time, solvation effects have to be included for reactions in solution. It is beyond the scope of this book to deal with these matters, but excellent progress is being made currently in computational chemistry by combining statistical thermodynamics and quantum chemistry in tackling what just a few years ago seemed intractable difficulties.

The effect of transferring an overall chemical reaction from the gas phase into a solvent is the difference between the effects upon the final and initial states, i.e. upon reactant and product molecules. The effect of this transfer upon a rate constant corresponds to the difference between the effects upon the transition state and initial state. For all states, enthalpy and entropy effects are involved.

Problems

1.1 Using Fig. 1.18 as a template, sketch a fully annotated reaction map for the following S_N2 reaction.

$$PhCH_2-Br \ + \ PhO^- \ \rightarrow \ PhCH_2-OPh \ + \ Br^-$$

By modifications to the reaction map, describe the effects upon the reaction mechanism of (i) electron-withdrawing substituents, and (ii) electron-donating substituents, in the two phenyl rings.

If a structural modification to a reactant leads to parallel *and* perpendicular effects, the overall effect upon the transition structure is the resultant of the two.

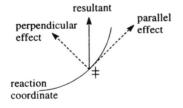

1.2 Acyl transfer reactions usually occur by step-wise addition-elimination mechanisms, eqn 1.4, Fig. 1.19(b). What sorts of groups (R, X$^-$, and Y$^-$) would lead to mechanisms becoming either concerted or even step-wise via an acylium intermediate? Sketch reaction maps for these alternatives.

1.3 Construct a fully annotated reaction map template for base-induced α,β-elimination from RCH_2-CHXR' using β-proton abstraction and departure of the nucleofuge, X$^-$, as the component reaction coordinates.

(a) Include E1, E2, and E1CB mechanisms.

(b) For substrates $PhCH_2CH_2X$ and CH_3CHXPh, how will (i) the nature of X, (ii) substituents in the phenyl groups, and (iii) the strength of the base determine the mechanism?

1.4 Draw annotated reaction map templates for the following.

(a) Cyclobutanone $\rightarrow$ $CH_2=CH_2$ + $CH_2=C=O$

(b) The retro-cycloaddition of eqn 1.5 to give propenal and butadiene.

For both, include reaction paths for concerted and step-wise mechanisms. What sort of substituents in butadiene would promote a step-wise mechanism through a zwitter-ion intermediate for the cycloaddition direction of reaction (b)?

2   Molecular mechanisms and reaction rates

2.1 Introduction

The principal technique for the investigation of mechanisms of chemical reactions is chemical kinetics. It is really a group of techniques whose common feature is that they involve measurements of rates of reactions. Some aspects will be illustrated here and others in later parts of the book.

2.2 Rate laws, rate constants, and mechanism

A simple bimolecular mechanism between like or unlike molecules will lead to a second-order rate law. For example, in solution, the reaction

$$Et_3N \;+\; EtBr \;\rightarrow\; Et_4N^+\, Br^-$$

has the rate law

$$-d[Et_3N]/dt \;=\; -d[EtBr]/dt \;=\; k\,[Et_3N]\,[EtBr]\,;$$

and, in the gas phase, the reaction

has the rate law

$$-d[C_4H_6]/dt \;=\; k\,[C_4H_6]^2\,.$$

In these and other comparable reactions, the second-order rate constants, k, usually have units $dm^3\ mol^{-1}\ s^{-1}$.

Correspondingly, simple unimolecular mechanisms lead to first-order rate laws, for example,

$$-d[\textit{cis}\text{-but-2-ene}]/dt \;=\; k\,[\textit{cis}\text{-but-2-ene}]\,,$$

and the first-order rate constants of such reactions, k, usually have units s^{-1}.

For each of these reactions, it is the rate constant which characterises the kinetics of the transformation under specified conditions of temperature, solvent (for reactions in solution), or pressure (for reactions in the gas phase). The *differential* terms on the left in each case are the *rates* at which

Unlike the rate constant, the rate of a chemical reaction usually slows down as the reaction proceeds and reactants are used up.

The frequency of genuine trimolecular collisions between molecules is so low that trimolecular elementary steps make virtually no contribution to chemical reactivity.

concentrations of reactants decrease with time. For reactions at constant volume, these are directly proportional to the rate of reaction, and are commonly but imprecisely taken actually to be the rate of reaction.

Many reactions of organic compounds, both in solution and in the gas phase, do have first- or second-order rate laws which follow from the simplicity of their mechanisms as exemplified by the above. The converse is not true, however; establishing that a reaction's rate law is either first or second order is not sufficient evidence that its mechanism is either unimolecular or bimolecular, respectively. Reactions with quite complicated mechanisms are known which, for rather particular reasons, have deceptively simple rate laws. However, it does follow that a complicated rate law is the consequence of a complex mechanism, but this must ultimately comprise interconnected simple elementary steps which individually are either unimolecular or bimolecular.

As indicated above, simple unimolecular and bimolecular mechanisms necessarily lead to first-order and second-order rate laws. Usually, though, we are dealing with a reaction whose mechanism we wish to understand, i.e. in practice, we start with a reaction, not a mechanism! The strategy in the elucidation of mechanism, therefore, is to carry out exploratory investigations, produce a tentative mechanism, predict some consequences from the mechanism, then investigate them. The initial evidence will be product analysis as it is clearly essential that we know what the reaction is before we can investigate its mechanism. A provisional mechanism which accounts for the products is then proposed from which a rate law can be predicted and investigated. If the predicted rate law is not observed experimentally, however, the mechanism is defective and needs to be modified or replaced in the light of the experimental rate law. At the stage when the rate law required by the mechanism is identical with that observed, then some confidence in the mechanism is justifiable. More stringent and subtle kinetics and product analysis tests may next be applied.

2.3 Simple transition state theory

The free energy of activation and reaction profiles

For a simple unimolecular reaction in which reactant molecule A proceeds via the activated complex $A^{\ddagger}$ as in eqn 2.1,

$$A \; \underset{(\longleftarrow)}{\xrightarrow{\hspace{2cm}}} \; A^{\ddagger} \; \longrightarrow \; \text{Product} \qquad (2.1)$$

the relationship between the experimental first-order rate constant, k, and the molar free energy of activation, $\Delta G^{\ddagger}$, is given by eqn 2.2 where all symbols have their usual meanings.

$$k = \frac{k_B T}{h} \cdot e^{-\Delta G^{\ddagger}/RT} \qquad (2.2)$$

It follows that the rate constant and the free energy of activation are alternative temperature-dependent parameters which describe what we often loosely call the rate of a reaction. A high value for k and a small $\Delta G^{\ddagger}$ correspond to a fast reaction, and a low value for k and a large $\Delta G^{\ddagger}$ correspond to a slow reaction. The reaction may be described diagrammatically as shown in Fig. 2.1.

A transmission coefficient, κ, is occasionally included in eqn 2.2. This is almost invariably assumed to be unity so, for simplicity, we have left it out.

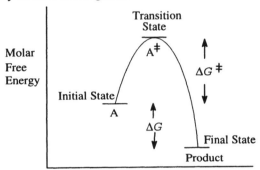

Fig. 2.1 Molar free energy profile for the unimolecular first-order reaction of eqn 2.1

An analogous free energy profile is shown in Fig. 2.2 for the bimolecular reaction of eqn 2.3.

$$A + B \xrightleftharpoons{\quad(\longleftarrow)\quad} AB^{\ddagger} \longrightarrow \text{Product} \qquad (2.3)$$

Strictly, the molar free energy terms of eqn 2.2 and in profiles such as Fig. 2.1 are *standard* molar free energies, and should be labelled $\Delta G^{\ominus\ddagger}$ and $\Delta G^{\ominus}$, i.e. they refer to molar free energies of substances in a specified standard state. In kinetics, especially for reactions in solution, the material at 1 mol dm^{-3} under 1 atm pressure in a specified solvent and at a specified temperature is the most convenient standard state. Here and subsequently, we omit the word 'standard' for convenience and, in accord with common practice, dispense with the standard state superscript.

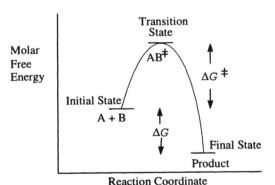

Fig. 2.2 Molar free energy profile for the bimolecular second-order reaction of eqn 2.3

The profiles in Figs 2.1 and 2.2 include molar free energy plotted against the reaction coordinate; initial state, transition state, and final state are indicated and represented by horizontal lines. The diagrams also include the reactant molecules and activated complexes (and could have included product molecules). It is to the interconversion of these molecular species (rather than the thermodynamic states) that the continuous curved line representing molecular potential energy corresponds. The distinction between the molecular species and the thermodynamic states is crucial. The former are real if the mechanism is correct and the latter are hypothetical.

Here also, eqn 2.2 describes the relationship between the (standard) molar free energy of activation and the experimental rate constant. But there is a significant difference. When eqn 2.2 is used to calculate a reaction's free energy of activation, the experimental second-order rate constant must have units dm^3 mol^{-1} s^{-1} in order that the free energy of activation corresponds to a standard state of 1 mol dm^{-3}. This is not an issue for first-order rate constants since their units do not include a concentration term.

One of the postulates of transition state theory is that the reactant molecules are in thermodynamic equilibrium with activated complex molecules. So, although the molecular species at the maximum in reaction profiles such as Figs 2.1 and 2.2 are exceedingly short-lived compared with the time-scale of the chemical reaction, they are regarded as being in thermal equilibrium with their surroundings. It follows, therefore, that we can associate an equilibrium constant, $K^{\ddagger}$, with this postulated equilibrium. Assuming further that the normal relationships of classical thermodynamics apply, this will be related to the free energy of activation, $\Delta G^{\ddagger}$, by the equation

> This postulated equilibrium between initial state and transition state implies nothing about other possible transformations of the activated complex molecules.

$$\Delta G^{\ddagger} = -RT \ln K^{\ddagger}.$$

Formally, the molar free energy of activation is given by

$$\Delta G^{\ddagger} = \Delta_f G(\text{activated complex}) - \Sigma \Delta_f G(\text{reactants})$$

where $\Delta_f G$ means the molar free energy of formation under the conditions of the reaction (temperature, solvent, pressure, etc.).

> Here and throughout the book, we use ln to mean ln$_e$, i.e. logarithms to the base e (sometimes called natural logarithms).

Enthalpy and entropy of activation

The free energy of activation may be resolved into enthalpy and entropy components by the equation

$$\Delta G^{\ddagger} = \Delta H^{\ddagger} - T \Delta S^{\ddagger} \tag{2.4}$$

where $\Delta H^{\ddagger}$ = (standard) molar enthalpy of activation, and
$\Delta S^{\ddagger}$ = (standard) molar entropy of activation.

However, although $\Delta G^{\ddagger}$ at a particular temperature may be calculated from the rate constant using eqn 2.2, $\Delta H^{\ddagger}$ and $\Delta S^{\ddagger}$ cannot.

By substituting eqn 2.4 into eqn 2.2, we obtain

$$k_c = \frac{k_B T}{h} . e^{-(\Delta H^{\ddagger} - T\Delta S^{\ddagger})/RT}$$

where k_c now represents either a first-order rate constant, or a second-order rate constant with the proper concentration units (mol dm^{-3}).

This may be rewritten

$$k_c = \frac{k_B T}{h} . e^{\Delta S^{\ddagger}/R} . e^{-\Delta H^{\ddagger}/RT} \tag{2.5}$$

or $$\ln k_c = \ln\left(\frac{k_B}{h}\right) + \ln T + \frac{\Delta S^{\ddagger}}{R} - \frac{\Delta H^{\ddagger}}{RT}.$$

Differentiation with respect to temperature gives

$$\frac{d(\ln k_c)}{dT} = \frac{1}{T} + \frac{1}{R} . \frac{d(\Delta S^{\ddagger})}{dT} - \frac{1}{RT} . \frac{d(\Delta H^{\ddagger})}{dT} + \frac{\Delta H^{\ddagger}}{RT^2}$$

but
$$\frac{d(\Delta S^{\ddagger})}{dT} = \frac{1}{T} \cdot \frac{d(\Delta H^{\ddagger})}{dT}$$

so
$$\frac{d(\ln k_c)}{dT} = \frac{\Delta H^{\ddagger} + RT}{RT^2}. \tag{2.6}$$

The empirical Arrhenius equation also describes the temperature dependence of the rate constant of a chemical reaction, and its logarithmic form is

$$\ln k_c = \ln A - \frac{E_a}{RT}$$

where
A = the Arrhenius pre-exponential factor,
E_a = the Arrhenius activation energy, and
T = the absolute temperature.

If we assume that A and E_a are constants, and differentiate this equation with respect to temperature, we obtain

$$\frac{d(\ln k_c)}{dT} = \frac{E_a}{RT^2}$$

Linearity in an Arrhenius plot of $\ln k_c$ against $(T)^{-1}$ is evidence of the temperature independence of A and E_a.

which, following comparison with eqn 2.6, establishes the relationship between E_a and $\Delta H^{\ddagger}$ shown in eqn 2.7.

$$E_a = \Delta H^{\ddagger} + RT. \tag{2.7}$$

Numerically, RT is usually quite small compared with E_a.

We have, therefore, a ready method of determining $\Delta H^{\ddagger}$ from rate constant measurements at different temperatures; we simply use E_a obtained from the gradient of an Arrhenius plot of $\ln(k_c)$ against $(T)^{-1}$ and eqn 2.7. For T in eqn 2.7, we use the mean temperature of the Arrhenius investigation.

Next, we substitute for $\Delta H^{\ddagger}$ in eqn 2.5 using eqn 2.7 to obtain

$$k_c = \frac{k_B T}{h} \cdot e^{\Delta S^{\ddagger}/R} \cdot e^{-(E_a - RT)/RT}$$

or
$$k_c = \frac{k_B T}{h} \cdot e^{\Delta S^{\ddagger}/R} \cdot e^{-E_a/RT} \cdot e$$

and comparison of this with the exponential form of the Arrhenius equation,

$$k_c = A. e^{-E_a/RT},$$

establishes that
$$A = \frac{k_B T}{h} \cdot e \cdot e^{\Delta S^{\ddagger}/R}. \tag{2.8}$$

We see, therefore, that an experimental value for $\Delta S^{\ddagger}$ may be obtained from the pre-exponential factor, A, determined from the intercept in an Arrhenius plot of $\ln k_c$ versus $(T)^{-1}$. In eqn 2.8, as in eqn 2.7, T is the mean temperature of the Arrhenius investigation and, if the reaction is other than first order, k_c values must have mol dm^{-3} concentration units.

It turns out that $\Delta H^{\ddagger}$ and $\Delta S^{\ddagger}$ are relatively insensitive to temperature.

The Eyring equation

In the above sections, we saw that it is possible to determine $\Delta H^{\ddagger}$ and $\Delta S^{\ddagger}$ from the Arrhenius parameters E_a and A, respectively. It is possible to determine them directly from the same experimental results by a slightly different approach. For most reactions in solution, this is the commonest and most convenient way to proceed.

Rewriting eqn 2.2 using k_c rather than k to remind us to use rate constants with the proper units

$$k_c = \frac{k_B T}{h} \cdot e^{-\Delta G^{\ddagger}/RT}$$

and substituting for $\Delta G^{\ddagger}$ we obtain, after a minor rearrangement,

$$\frac{k_c}{T} = \frac{k_B}{h} \cdot e^{-(\Delta H^{\ddagger} - T\Delta S^{\ddagger})/RT}$$

By separating the exponential term into two, taking logarithms, and rearranging, we obtain the equation

$$\ln\left(\frac{k_c}{T}\right) = \ln\left(\frac{k_B}{h}\right) + \frac{\Delta S^{\ddagger}}{R} - \frac{\Delta H^{\ddagger}}{RT}$$

which shows that the gradient is $-\Delta H^{\ddagger}/R$ in a graph of $\ln(k_c/T)$ against $(T)^{-1}$, and the intercept is $\ln(k_B/h) + \Delta S^{\ddagger}/R$.

2.4 The physical significance and mechanistic interpretation of activation parameters

Arrhenius activation energy, E_a

The activation energy is the parameter in the empirical Arrhenius equation which describes the temperature dependence of the rate constant. A large E_a corresponds to a rate constant which increases rapidly with temperature, and a small E_a corresponds to a rate constant which increases only slightly with temperature. However, E_a is related by eqn 2.7 above to the enthalpy of activation, $\Delta H^{\ddagger}$, a term whose origin lies in transition state theory and which is of considerable mechanistic significance (see below).

Note that $E_a = 0$ corresponds to a reaction whose rate constant is temperature-independent, and that, strictly, the value of E_a alone gives no information about the absolute value of the rate constant - only its temperature dependence.

Arrhenius pre-exponential factor, A

Formally, the pre-exponential factor in the Arrhenius equation is the hypothetical value of the rate constant as the temperature approaches infinity. It necessarily has the same units as the rate constant. Consequently, A values of different reactions may be legitimately compared only if they are of the same kinetic order. Although A originates in an empirical equation, we saw above in eqn 2.8 that it is related to the entropy of activation which is based in theory. As we shall see below, therefore, A indirectly provides mechanistic information via $\Delta S^{\ddagger}$.

Free energy of activation, $\Delta G^{\ddagger}$

The free energy of activation is a parameter which originates in transition state theory and provides the same information as the rate constant–they are interconvertible using eqn 2.2. Both quantify and express what we may otherwise state qualitatively by saying, for example, that a reaction is fast or slow, and neither provides more information or mechanistic insight than the other. The rate constant has the advantage that its units indicate the rate law of the reaction (for example, first order or second order), and warns us against inappropriately comparing reactions of different order. On the other hand, the free energy of activation allows us to represent the reaction diagrammatically in a reaction profile, and offers a *pseudo*-thermodynamic analysis of the kinetics in terms of enthalpy and entropy of activation.

Enthalpy of activation, $\Delta H^{\ddagger}$

The enthalpy of activation is the component of the free energy of activation which is the enthalpy change as one mole of initial state becomes one mole of transition state. For gas-phase reactions, $\Delta H^{\ddagger}$ is the molar enthalpy change due to rebonding as one mole of reactant molecules become one mole of activated complex molecules, i.e. a theoretical process,

$$\Delta H^{\ddagger} = \Delta_f H(\text{activated complex}) - \Sigma \Delta_f H(\text{reactant molecules}), \quad (2.9)$$

where $\Delta_f H$ = molar enthalpy of formation.

Gas-phase homolysis reaction. In the gas-phase homolysis of ethane to give two methyl radicals,

$$CH_3-CH_3 \rightarrow 2\,CH_3^{\cdot} \quad (\rightarrow \text{further reaction}), \quad (2.10)$$

the experimental value of $\Delta H^{\ddagger} = 361\ \text{kJ mol}^{-1}$ (obtained from $E_a = 368\ \text{kJ mol}^{-1}$ over the temperature range 823 - 893 K using eqn 2.7) is virtually the same as the C–C bond dissociation enthalpy. The C–C bond, therefore, is almost completely broken in the transition state for homolysis and (according to this evidence) the activated complex comprises a pair of loosely associated methyl radicals.

Gas-phase isomerisation. Equation 2.9 could be reformulated as eqn 2.11 where DH = molar bond dissociation enthalpy.

$$\Delta H^{\ddagger} = \Sigma DH(\text{bonds broken}) - \Sigma DH(\text{bonds formed}) \quad (2.11)$$

This shows more clearly that a low enthalpy of activation may be anticipated if the formation of new bonds is concerted with the cleavage of old ones in the formation of the transition state.

For the gas-phase isomerisation in eqn 2.12, $E_a = 123\ \text{kJ mol}^{-1}$ from rate constant measurements made over the temperature range 416 - 467 K hence, by eqn 2.7, $\Delta H^{\ddagger} = 119\ \text{kJ mol}^{-1}$.

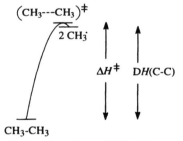

Fig. 2.3 Enthalpy profile for the homolysis of ethane
The closeness of the transition structure to the product in energy leads to the deduction that, *structurally*, it is very product-like (see Hammond effect, p. 11).

$$\text{(2.12)}$$

This value is much smaller than typical bond dissociation enthalpies and indicates that bond breaking and bond making are concerted when reactant molecules are transformed into activated complex molecules as illustrated in eqn 2.12 where the broken lines represent partial bonds.

Bimolecular substitution reaction in solution. The rate of the substitution reaction in eqn 2.13 was investigated in acetone as solvent. Measurement of the second-order rate constant at several temperatures around 45 °C gave an Arrhenius plot, and $E_a = 73$ kJ mol^{-1} hence $\Delta H^{\ddagger} = 71$ kJ mol^{-1}.

$$C_2H_5\text{-Br} + Cl^- \rightarrow C_2H_5\text{-Cl} + Br^- \qquad (2.13)$$

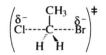

Fig. 2.4 Activated complex in the reaction of eqn 2.13

This rather low value is due to the partial formation of the C–Cl bond being concerted with the cleavage of the C–Br bond, i.e. the enthalpy cost of partially breaking one bond is offset by the partial formation of another. The activated complex in this S_N2 mechanism is represented in Fig. 2.4.

Additionally, however, as for all reactions in solution, the enthalpy of solvation of the initial state and of the transition state have to be taken into account. Chloride, being a relatively small anion of high charge density is strongly solvated whereas the activated complex shown in Fig. 2.4 has the same negative charge distributed over a much larger species. Consequently, the activated complex has a smaller polarising effect upon solvent molecules and so is less strongly solvated. Its formation, therefore, requires some desolvation of chloride which involves an enthalpy cost, and this is included in the enthalpy of activation. However, the liberation of solvent molecules is entropically very favourable as we shall see in the next section.

Unimolecular substitution/elimination reaction in solution. The importance of solvation in a unimolecular solvolysis is illustrated by the reaction of *tert*-butyl chloride (2-methyl-2-chloropropane) in methanoic acid.

$$(2.14)$$

In eqn 2.14, which is kinetically first order, the two products are formed from the *tert*-butyl carbenium ion intermediate either by solvent capture followed by proton loss (S_N1) or by direct proton donation to the solvent (E1). A simplified description of this type of $S_N1/E1$ mechanism is shown in eqn 2.15 and includes a rate-determining ionization followed by partitioning of the carbenium ion between the alternative parallel product-forming steps.

$$R\text{—}X \longrightarrow \left(\overset{\delta+}{R}\text{------}\overset{\delta-}{X}\right)^{\ddagger} \longrightarrow \left[R^+ \; X^-\right] \text{-----}\rightarrow \text{Products} \qquad (2.15)$$

The enthalpy of activation of this reaction is 88 kJ mol^{-1}. Whilst this is appreciably greater than values for typical S_N2 reactions, it is still much smaller than the bond dissociation enthalpy of a C–Cl bond (ca. 350 kJ mol^{-1} in the gas phase). There are two possible reasons for this. First, the bond in the activated complex included in eqn 2.15 is not completely broken–there is still vestigial bonding between the carbon and the chlorine in the transition state. Secondly, the dipolar activated complex will be much more strongly solvated than a relatively nonpolar *tert*-butyl chloride reactant molecule by the polar methanoic acid molecules through hydrogen bonds and Lewis acid-base interactions. This greater solvation of the activated complex molecules relative to the reactant molecules reduces the enthalpy cost of the ionization in solution. And the better the solvent is able selectively to solvate the activated complex, the lower the enthalpy barrier as shown in Fig. 2.5.

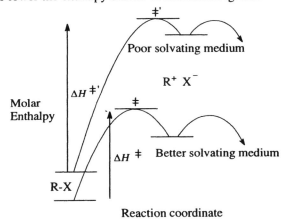

Fig. 2.5 Effect of solvation on the enthalpy of activation of ionization of neutral R-X

Fig. 2.5 describes the ionisation of neutral substrate R-X. In the gas phase, there would not be a minimum corresponding to the ion pair, $R^+ X^-$. In the poor solvating medium, there is a shallow minimum, and the barrier to its formation is $\Delta H^{\ddagger'}$. In the better solvating medium, both reactant and intermediate are stabilised, but the latter more than the former, and the activated complex is between the two. The consequence is that the enthalpy barrier to ionisation, $\Delta H^{\ddagger}$, is appreciably smaller in the better solvating medium.

There are two corollaries of this interpretation. First, if the solvent does not have Lewis base sites to interact with the developing positive charge on the α-carbon, and does not have weakly acidic hydrogens to form hydrogen bonds with the developing nucleofuge, then there will be no selective stabilisation of the activated complex and the enthalpy cost of ionization will be high. This is observed. The enthalpy cost of ionisation in solvents such as hydrocarbons (which will solvate the nonpolar reactant molecules better than polar activated complex molecules) is prohibitively high, and S_N1 reactions do not occur in such solvents. Secondly, the enhanced solvation of the activated complex molecules in a polar solvent corresponds to decreased rotational and translational freedom of the solvent molecules in the transition state. This should contribute adversely to the entropy of activation which is also observed as we shall see in the next section.

Note that reduction of the energy difference between reactant and ion-pair by transferring the reaction to a better ionizing medium moves the transition structure for ionization towards the reactant; this is a parallel (Hammond) effect, see chapter 1.

Entropy of activation, $\Delta S^{\ddagger}$

The entropy of activation is the non-enthalpic component of the free energy of activation. Like $\Delta H^{\ddagger}$, it is approximately temperature independent but, unlike $\Delta H^{\ddagger}$, it is difficult to interpret in general terms. Particular values admit quite different interpretations according to the natures of the reactions. However, as its relationship with the pre-exponential factor indicates (eqn 2.8), it is not

Whereas $\Delta H^{\ddagger}$ is independent of the units of the rate constants used in its determination, this is not true for $\Delta S^{\ddagger}$.

$\Delta S^{\ddagger}$ = S(activated complex)
  − S(reactant)

legitimate to compare $\Delta S^{\ddagger}$ values for reactions of different kinetic orders. Consequently, in this respect also, it differs from $\Delta H^{\ddagger}$.

Gas-phase homolysis reaction. For the dissociation of ethane into methyl radicals, eqn 2.10 above, $A = 5.0 \times 10^{16}$ s^{-1} which leads by eqn 2.8 to $\Delta S^{\ddagger} = 58$ J K^{-1} mol^{-1}. This is a large positive result–exactly what would be expected if the activated complex comprises a very loosely bonded pair of methyl groups, i.e. the mechanism proposed in Fig. 2.3 on the basis of the enthalpy of activation evidence.

Gas-phase isomerisation. In contrast, $\Delta S^{\ddagger} = -32$ J K^{-1} mol^{-1} (calculated from $A = 5.4 \times 10^{11}$ s^{-1}) for the gas-phase isomerisation of eqn 2.12. This negative result, i.e. a loss of entropy as one mole of initial state gives one mole of transition state, is qualitatively what would be expected if a conformationally mobile reactant molecule has to assume a restricted conformation in order that rebonding can occur in the activated complex shown in eqn 2.12. This entropy of activation result, therefore, is also fully in accord with the mechanism proposed from the enthalpy of activation evidence–a unimolecular mechanism through a cyclic activated complex involving concerted bond breaking and making.

$\Delta S^{\ddagger}$ = S(activated complex)solvent
  − Σ S(reactants)solvent

Note the units of A; the reaction of eqn 2.13 is second order and A has the same units as the rate constant.

Bimolecular substitution reaction in solution. Any bimolecular mechanism involves a loss of translational entropy in the formation of the transition state because each activated complex molecule is derived from two reactant molecules (or ions). In the bimolecular substitution reaction of eqn 2.13 (p. 22), the reactants are an ethyl bromide molecule and a chloride anion. From $A = 7.9 \times 10^{8}$ dm^{3} mol^{-1} s^{-1}, we calculate $\Delta S^{\ddagger} = -75$ J K^{-1} mol^{-1}. In fact, the purely translational contribution due to the formation of the activated complex shown in Fig. 2.4 from reactants may be calculated and is even more negative than -75 J K^{-1} mol^{-1}. There must be another feature of this reaction, therefore, which makes a positive contribution to the overall experimental $\Delta S^{\ddagger}$, and this is the change in solvation associated with the formation of the transition state. When considering the enthalpy of activation of this reaction above, we saw that the activated complex shown in Fig. 2.4 is less strongly solvated than chloride. This shedding of solvent molecules from around the chloride makes a positive contribution to the overall entropy of activation, hence the experimental value for $\Delta S^{\ddagger}$ is less negative than expected from just the reaction's bimolecularity.

In other substitution reactions in solution, solvation effects can amplify the translational contribution of the bimolecularity. In the quaternisation reaction of eqn 2.16, two neutral molecules form a tetra-alkylammonium iodide, and the activated complex involves some degree of dipolarity.

$$Et_3\ddot{N} + \text{Et-I} \longrightarrow \left(Et_3\overset{\delta+}{N}\text{---}\underset{\underset{CH_3}{|}}{CH_2}\text{--}\overset{\delta-}{I} \right)^{\ddagger} \longrightarrow Et_4N^+ \ I^- \qquad (2.16)$$

In this reaction, the activated complex molecules are more strongly solvated than the relatively nonpolar reactant molecules which leads to a strong negative contribution to the entropy of activation. Experimental values range between -160 and -200 J K^{-1} mol^{-1} depending upon the solvent. Correspondingly, the enthalpies of activation are very low, $\Delta H^{\ddagger} = 46 - 50$ kJ mol^{-1} depending on the solvent, as the enthalpic cost of forming the activated complex from reactant molecules is offset by the enthalpy of intermolecular interactions between the solvating molecules and the activated complex.

Unimolecular substitution/elimination reactions in solution. The initial rate-limiting step of the S$_N$1/E1 mechanism of eqn 2.15 involves cleavage of the bond between the nucleofuge and the α-carbon. Formation of the activated complex, therefore, involves appreciable loosening of this bond and, as in the homolysis of ethane in eqn 2.10 and Fig. 2.3, would lead to a positive entropy of activation if nothing else were involved. However, there are two significant differences between the reactions of eqns 2.10 and 2.15. The former is in the gas phase and is a homolysis whereas the latter is in solution and is a heterolysis. Consequently, the formation of a dipolar activated complex from a relatively nonpolar reactant in eqn 2.15 leads to increased solvation. This corresponds to a decrease in the translational and rotational freedom of solvent molecules, i.e. a negative contribution to the overall entropy of activation. Typically, $\Delta S^{\ddagger}$ values for S$_N$1/E1 reactions of neutral compounds as in eqn 2.15 are close to zero (-7 J K^{-1} mol^{-1} for the formolysis of *tert*-butyl chloride, eqn 2.14), so it appears that the gain in entropy due to the bond loosening in the heterolysis is just about balanced by the loss of entropy due to the attendant increase in solvation in the transition state.

$\Delta S^{\ddagger} = S$(activated complex)solvent
$- S$(reactant)solvent

Formolysis is solvolysis in formic acid, i.e. methanoic acid.

$$ (2.17) $$

Anomalous effects are observed for S$_N$1/E1 reactions in water. The entropy of activation for hydrolysis of *tert*-butyl chloride, eqn 2.17, is strongly positive ($\Delta S^{\ddagger} = 51$ J K^{-1} mol^{-1}). In addition to the effects described above, formation of the solvated dipolar activated complex disrupts the three-dimensional hydrogen-bonded structure of water. This makes an appreciable positive contribution to the overall entropy of activation.

Problems

2.1 The temperature dependence of the first-order rate constant of the gas-phase isomerisation of cyclobutene is given by the following equation.

$$ k = 1.2 \times 10^{13} \exp(-16356/T) \text{ s}^{-1}. $$

(a) Comparing this with the exponential form of the Arrhenius equation on p. 19, calculate E_a, $\Delta H^{\ddagger}$, and $\Delta S^{\ddagger}$, and comment on their values.

(b) At 150°C, the experimental rate constant is $2.0 \times 10^{-4}\,s^{-1}$; does this fit the above expression?

(c) Calculate $\Delta G^{\ddagger}$ from the value of k given in part (b); how does your value compare with that calculated from the results of part (a) at 150°C?

2.2 Calculate $\Delta H^{\ddagger}$ and $\Delta S^{\ddagger}$ from the Arrhenius parameters given for the following two reactions, and comment upon their relative values. How do you account for the difference in the $\Delta S^{\ddagger}$ values?

We use log to mean decadic logarithms, $\log_{10}$, i.e. logarithms to the base 10.

$\xrightarrow{\text{482 - 543 K}}$ 3 CH$_3$CHO ;

$\log(A/s^{-1}) = 15.1$, $E_a = 185$ kJ mol^{-1}

CH_3-C ... $O-CH(CH_3)_2$ $\xrightarrow{\text{715 - 801 K}}$ $CH_3CO_2H + CH_3CH=CH_2$;

$\log(A/s^{-1}) = 13.0$, $E_a = 188$ kJ mol^{-1}

Calculate the rate constants for these two reactions at a common temperature of 600K, and comment upon their relative values.

2.3 Calculate $\Delta H^{\ddagger}$ and $\Delta S^{\ddagger}$ from the Arrhenius parameters given for the following two reactions. Why are the $\Delta H^{\ddagger}$ values so different whereas the $\Delta S^{\ddagger}$ values are so similar?

$\xrightarrow{\text{723 - 823 K}}$; $\log(A/s^{-1}) = 13.0$; $E_a = 272$ kJ mol^{-1}

$\xrightarrow{\text{553 - 615 K}}$; $\log(A/s^{-1}) = 12.8$; $E_a = 179$ kJ mol^{-1}

2.4 The decomposition of azomethane, Me-N=N-Me, in the gas phase occurs by rate-limiting homolysis of a C–N bond; for this reaction, $\Delta H^{\ddagger} = 227$ kJ mol^{-1} and $\Delta S^{\ddagger} = 73$ J K^{-1} mol^{-1}.

Azo-*iso*-butyronitrile, Me$_2$C(CN)-N=N-C(CN)Me$_2$, decomposes by a similar mechanism but at much lower temperatures in an inert solvent; for this reaction, $\Delta H^{\ddagger} = 131$ kJ mol^{-1} and $\Delta S^{\ddagger} = 51$ J K^{-1} mol^{-1}.

How do you account for (a) the appreciable difference between the $\Delta H^{\ddagger}$ values, and (b) the substantially positive $\Delta S^{\ddagger}$ values?

3   Correlation analysis: quantitative relationships between molecular structure and chemical reactivity

3.1 Introduction

An appreciation of the relationship between structure and reactivity is central to modern organic chemistry, and qualitative notions are introduced from the earliest stages. Consequently, organic chemists can usually predict whether replacement of a hydrogen by a methyl, for example, or a chlorine by a fluorine in a compound will be rate-increasing or rate-decreasing in a particular reaction. Correlation analysis is a technique to facilitate quantification of such predictions.

Correlation analysis deals with quantitative effects of structural changes upon equilibrium constants as well as rate constants.

The commonest procedure is for reactants which include an aromatic ring, Fig. 3.1, since they allow the introduction of a wide range of substituents, X, remote from the reaction site, Z. Here, we have a chemical reaction of a compound $X-C_6H_4-Z$ in which the group Z is transformed into group W, and we may expect that substituents X will have an effect upon this reaction.

Substituents X are generally meta or para to Z and not, as we shall discuss later, ortho.

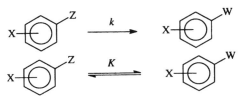

Fig. 3.1 Substituents X and reaction sites Z in reactions of aromatic compounds

But how do we correlate the effect upon either the rate or the equilibrium for the reaction at Z with the nature of the substituent X? We know that k and K quantify the rate and the equilibrium of the reaction, but what about the substituents? How do we put a number on whether X is F, CH_3, or any other substituent? We cannot, so we need a parameter which provides a numerical value for a *property* of the substituent. The property that was introduced for

The reaction site in Z may be bonded directly to the benzene ring, e.g. Z = -CO$_2$R, or it may be attached to the benzene ring by another group, e.g. Z = -CH$_2$-CO$_2$R.

this purpose is the effect of the substituent upon the dissociation of benzoic acid in water at 25°C.

Correlation analysis, therefore, involves comparison of the effect (relative to hydrogen) of a substituent in a compound in the reaction under scrutiny with the effect of that same substituent (relative to hydrogen) upon a standard reaction. In other words, structural changes to a reactant are described in terms of replacement of hydrogen by a substituent X, and a parameter is ascribed to each substituent which quantifies its effect relative to hydrogen upon a standard reaction–the dissociation of benzoic acid. Consideration first of the standard reaction, then of real applications, should clarify what is not at all easy to apprehend in general.

3.2 Effect of substituents upon the acidity of benzoic acid in water at 25°C–definition of the substituent parameter, σ

Equation 3.1 represents the dissociation of a family of substituted benzoic acids in water at 25°C. For each member of the family, there is a dissociation constant defined in the conventional way, eqn 3.2, determinable experimentally, and often expressed logarithmically, i.e. $pK_a = -\log K_a$. A few representative pK_a values are given in Table 3.1.

Note that we are not dealing with a single reaction–we are necessarily dealing with a family (or series) of reactions, all of the same type at Z. There will be as many individual reactants and hence reactions in the family as there are alternative substituents X.

As usual, mole fraction is used as the activity scale for water in aqueous solutions so its activity is effectively unity for a dilute solution of any solute and hence does not appear in eqn 3.2.

$$X\text{–C}_6H_4\text{–CO}_2H + H_2O \underset{25°C}{\overset{H_2O}{\rightleftharpoons}} X\text{–C}_6H_4\text{–CO}_2^- + H_3O^+ \tag{3.1}$$

$$\text{or}\quad AH + H_2O \rightleftharpoons A^- + H_3O^+$$

$$K_a = \frac{[A^-][H_3O^+]}{[AH]} \tag{3.2}$$

Substituent X	$pK_a(XC_6H_4CO_2H)$	σ_X
H	4.20	0
3-OMe	4.09	0.11
3-F	3.86	0.34
3-NO$_2$	3.49	0.71
4-NO$_2$	3.42	0.78
4-Me	4.37	-0.17
4-OMe	4.48	-0.28

Table 3.1 pK_a values of some substituted benzoic acids in water at 25°C, and σ-values of the substituents

σ_X means the σ parameter of the substituent X.

Each member of the family of benzoic acids is distinguished by its substituent, X, and a parameter σ_X is defined by eqn 3.3. We see, therefore, that σ is a measure of the ability of the substituent X to modify the acid strength of benzoic acid, and a few values are included in Table 3.1. A positive σ-value indicates an acid-strengthening substituent, and this invariably corresponds to an electron-attracting effect; a negative σ-value indicates an acid-weakening substituent, usually one with an electron-donating effect.

$$\sigma_X = pK_a(C_6H_5CO_2H) - pK_a(XC_6H_4CO_2H) \qquad (3.3)$$

The effect of replacing a hydrogen meta or para to the carboxyl group of benzoic acid by an electron-withdrawing substituent, e.g. nitro, upon the standard free energy for the dissociation of benzoic acid is illustrated in Fig. 3.2. The familiar relationship between the dissociation constant and the standard free energy of the reaction is given in eqn 3.4.

σ_X is a substituent parameter and, once determined by measurement of the acid dissociation constant of the X-substituted benzoic acid, may be used in any other appropriate context. Note that the σ value of a substituent, e.g. nitro in Table 3.1, depends upon whether the group is meta or para to the carboxyl group.

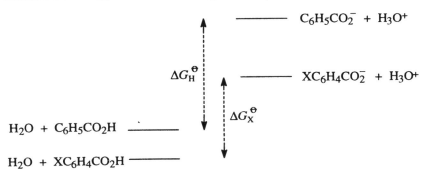

Fig. 3.2 Standard free energy diagram showing the effect of an electron-withdrawing substituent X upon the dissociation of benzoic acid in water

In these dissociations in water, hydronium ion is produced in both, and water as base on the left hand side is also common to both. Consequently, the difference between the two reactions is due wholly to the differential effect of the substituent upon the undissociated acid and its conjugate base, the anion.

$$-\Delta G^{\circ} = RT.\ln K_a \qquad (3.4)$$

Benzoic acid is only a weak acid in water ($K_a = 6.3 \times 10^{-5}$ at 25°C), and the standard molar free energy of reaction by eqn 3.4 is appreciably positive (+24 kJ mol^{-1}). Introduction of an electron-withdrawing substituent, e.g. 3-NO$_2$, has an effect upon both the undissociated acid and its anion. But the effect upon the anion is larger as illustrated in Fig. 3.2 with the consequence that the standard free energy of dissociation of 3-nitrobenzoic acid is less positive (+20 kJ mol^{-1}). In other words, the dissociation is less unfavourable, and the equilibrium constant somewhat larger ($K_a = 3.2 \times 10^{-4}$).

The adverse (positive) ΔG° for the dissociation of benzoic acid in water at 25°C is the result of $\Delta H^{\circ} \sim 0$ (as for carboxylic acids in general) and an appreciably negative ΔS° (-78.2 J K^{-1} mol^{-1}). The effect of substituents upon ΔG° is almost entirely due to solvation effects upon ΔS° and not upon ΔH°.

3.3 Logarithmic correlation of a family of equilibrium constants against σ–definition of the reaction parameter, ρ

Now that we have defined a substituent parameter, σ, we shall use it to correlate, for example, the acid strengths of substituted phenylethanoic acids to demonstrate a general strategy. In Fig. 3.3, we have a graph of logK_a for representative members of a series of X-substituted phenylethanoic acids against the σ values of the substituents. The graph is linear and the gradient, usually given the symbol ρ, is 0.49.

The relationship between logK_a for a series of X-substituted phenylethanoic acids and the σ values of the substituents may also be expressed mathematically as follows where ρ is the constant of proportionality, i.e. the gradient in the graph in Fig. 3.3.

$$\log K_a(XC_6H_4CH_2CO_2H) \propto \sigma_X$$

$$\log K_a(XC_6H_4CH_2CO_2H) = \rho.\sigma_X + \text{constant} \qquad (3.5)$$

For the parent compound, phenylethanoic acid itself, X = H and $\sigma_H = 0$. Consequently, the constant in eqn 3.5 may be evaluated,

$$\log K_a(C_6H_5CH_2CO_2H) = \text{constant},$$

and the above equation rewritten as

$$\log K_a(XC_6H_4CH_2CO_2H) = \rho.\sigma_X + \log K_a(C_6H_5CH_2CO_2H)$$

or

$$\log\left\{\frac{K_a(XC_6H_4CH_2CO_2H)}{K_a(C_6H_5CH_2CO_2H)}\right\} = \rho.\sigma_X \qquad (3.6)$$

where the value of ρ is 0.49.

The ρ-value obtained here is a parameter characteristic of this particular family of reactions–the dissociation of arylethanoic acids in water at 25°C. A different value would have been obtained, in principle, for the dissociation in water at a different temperature, or in a different solvent. Under the different reaction conditions, the different value for ρ would follow from different experimental $\log K_a$ results plotted against the same σ values.

In Fig. 3.3, we could have plotted pK_a against σ_X in which case the gradient would have been negative.

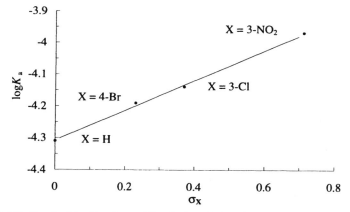

Fig. 3.3 Graph of $\log K_a$ values of X-substituted phenylethanoic acids against σ_X

If we were to construct another graph correlating $\log K_a$ values for X-substituted 3-phenylpropanoic acids ($XC_6H_4CH_2CH_2CO_2H$) against σ_X, we would again observe a linear plot. The gradient for this family of reactions (H_2O, 25°C) is $\rho = 0.21$. We see, therefore, that ρ is a parameter which characterises the particular family of reactions under specified experimental conditions.

3.4 The Hammett equation applied to equilibria

We saw above that the logarithms of equilibrium constants for two families of reactions, the dissociation of substituted phenylethanoic and 3-phenylpropanoic acids in water at 25°C, give linear correlations when plotted against substituent parameters defined by the dissociation of substituted benzoic acids, eqn 3.3. This is not too surprising since they are three rather similar reactions. However, it turns out that other types of equilibria involving aromatic compounds may also be correlated in just the same way, and we shall see examples later. Thus, for a general equilibrium involving aromatic reactants and products as in eqn 3.7

$$X-\bigotimes^{Z} \xrightleftharpoons{K} X-\bigotimes^{W} \qquad (3.7)$$

we may write by analogy with eqn 3.6

$$\log\left\{\frac{K^X}{K^H}\right\} = \rho\cdot\sigma_X \qquad (3.8)$$

where K^X is the equilibrium constant for the X-substituted member of the family of compounds defined in the conventional way,

K^H is the equilibrium constant for the parent member of the series, i.e. the compound with X = H,

σ_X is the substituent parameter for X defined in eqn 3.3,

and ρ is the reaction parameter specific to the family of reactions described by a chemical equation represented here by eqn 3.7.

Equation 3.8 is one form of the celebrated Hammett equation for equilibria.

The dissociation of substituted benzoic acids in water at 25°C–the standard reaction with ρ = 1

If we rewrite eqn 3.3 which defines σ parameters for substituents,

$$\sigma_X = \log K_a(XC_6H_4CO_2H) - \log K_a(C_6H_5CO_2H)$$

or

$$\log\left\{\frac{K_a(XC_6H_4CO_2H)}{K_a(C_6H_5CO_2H)}\right\} = \sigma_X \qquad (3.9)$$

we see it has the form of eqn 3.8 with $\rho = 1$. The dissociation of substituted benzoic acids in water at 25°C, which allows the assignment of σ-values to substituents, therefore, is also the standard reaction for which $\rho = 1$.

Determination of the ρ-value for a family of equilibria

As indicated above, the substituent constant σ is defined by eqn 3.3, so from the pK_a value of an X-substituted benzoic acid, the σ-value of X may be calculated. There are other indirect methods by which σ-values may be

With the Hammett equation for equilibria, substituents which increase equilibrium constants decrease positive free energies of reaction (or make negative free energies more negative).

The Hammett equation may also be written in the 'intercept' form, i.e.

$$\log K^X = \rho\cdot\sigma_X + \log K^o$$

in which the term for the parent compound ($\log K^X$ with X = H) is just one of the experimental terms on the left hand side, and $\log K^o$ is the intercept in the graph of $\log K^X$ against σ_X (approximately the same value as $\log K^H$).

determined, and σ-values for over 500 substituents are now known a few of which are shown in Table 3.2.

Substituent	σ_m	σ_p
NO_2	0.71	0.78
CN	0.61	0.70
CF_3	0.43	0.54
Br	0.39	0.23
CH_3CO	0.38	0.48
Cl	0.37	0.22
CHO	0.36	0.44
I	0.35	0.28
F	0.34	0.06
OH	0.13	-0.38
OCH_3	0.11	-0.28
Ph	0.05	0
H	0	0
CH_3	-0.06	-0.17
C_2H_5	-0.07	-0.15
NMe_2	-0.15	-0.63
NH_2	-0.16	-0.57

Table 3.2 σ parameters of some common substituents

In order to determine the ρ-value for a family of equilibria represented by eqn 3.7 above, it is necessary to measure the equilibrium constant K for, say, four or five members of the family. A graph is then drawn of either $\log(K^X/K^H)$ or simply of $\log K^X$ against the σ-values of the substituents; both may include the term for X = H, $\sigma_H = 0$. In the former case, the plot goes through the origin, but not in the latter; the gradient is the same, however, and equal to ρ for the family of reactions.

If the experimental data are inaccurate, the two plots corresponding to alternative statistical methods of analysis will not lead to identical results.

The sign and magnitude of ρ

Positive ρ-values. A chemical equilibrium which is affected in precisely the same way by substituents as is the dissociation of benzoic acid will also have a ρ-value equal to 1. If the effect of substituents is in the same sense (those which enhance the acidity of benzoic acid, i.e. electron-withdrawing substituents, lead to larger equilibrium constants) but the effect is smaller, then a positive ρ-value smaller than 1 will be observed (as in the dissociation of phenylethanoic and 3-phenylpropanoic acids). Another example is the acid dissociation of substituted benzylammonium ions in water at 25°C shown in eqn 3.10 for which ρ = 0.72.

The larger the positive ρ-value, the greater the sensitivity of the equilibrium to the introduction of polar substituents in the sense that the more the equilibrium constant will be increased by electron-withdrawing substituents.

Reactions which are affected in the same sense as the dissociation of benzoic acids (electron-withdrawing substituents, i.e. those with positive σ values, lead to increased equilibrium constants), but to a greater degree, will have positive ρ-values greater that 1. Examples are the dissociation of substituted benzoic acids in non-aqueous solvents such as ethanol ($\rho = 1.65$), dimethylformamide ($\rho = 2.36$), and dimethyl sulfoxide ($\rho = 2.48$).

$$X-\underset{}{\bigcirc}\text{-}CH_2\text{-}\overset{+}{N}H_3 + H_2O \rightleftharpoons X-\underset{}{\bigcirc}\text{-}CH_2\text{-}NH_2 + H_3O^+ \qquad (3.10)$$

ρ-Value of zero. If the introduction of polar substituents has no effect upon the equilibrium constant, i.e. if the Hammett plot looks like that in Fig. 3.4, then $\rho = 0$. This could be because the reaction site is too remote from the substituents as in the dissociation of $Ar(CH_2)_nCO_2H$ in water where n is greater than about 4. Alternatively, a ρ-value of about zero would be expected if there is no appreciable charge difference between reactant and product as in the trans-esterification of eqn 3.11, or in equilibria controlled by orbital symmetry considerations.

$$ArCO_2Et + MeOH \rightleftharpoons ArCO_2Me + EtOH \qquad (3.11)$$

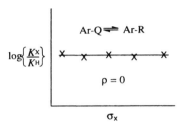

Fig. 3.4 Equilibria for which $\rho = 0$

Negative ρ-values. If the introduction into the reactant of substituents with positive σ-values (electron-withdrawing groups) reduces the equilibrium constant, then the plot of $\log(K^X/K^H)$ against the σ_X will have a negative gradient as in the sketch of Fig. 3.5, and the ρ-value will be negative.

Equation 3.10 above was written in the conventional manner with the proton donor on the left hand side, so the reaction is seen in terms of the ammonium cation as the acid. However, the reaction could be written differently with emphasis on the amine as a base, eqn 3.12. For this reaction, it is easily shown that $\rho = -0.72$.

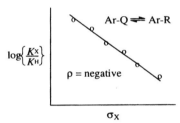

Fig. 3.5 Equilibria with negative ρ

$$X-\underset{}{\bigcirc}\text{-}CH_2\text{-}NH_2 + H_2O \rightleftharpoons X-\underset{}{\bigcirc}\text{-}CH_2\text{-}\overset{+}{N}H_3 + HO^- \qquad (3.12)$$

This illustrates that the chemical equilibrium must be explicit before the ρ-value is meaningful.

$$p\text{-}ClC_6H_4N_2^+ + ArSO_2^- \underset{25^oC}{\overset{CH_3OH}{\rightleftharpoons}} p\text{-}ClC_6H_4N=NSO_2Ar \qquad (3.13)$$

In the reaction between arenediazonium and arylsulfinate ions in methanol to give diazosulfones, substituents may be introduced into either the anion or cation. In the family of equilibria of eqn 3.13, electron-withdrawing substituents in the anion on the left relatively stabilise the left hand side. Such substituents decrease the equilibrium constant, and the reaction series has a negative ρ-value for differently substituted phenylsulfinate anions reacting with a common arenediazonium cation ($\rho = -2.07$).

Electron-withdrawing substituents (positive σ-values) introduced into the benzenediazonium ion will (relatively) destabilise the left hand side of eqn 3.13 and increase the equilibrium constant for reactions with a common

The magnitude of a ρ-value quantifies the sensitivity of an equilibrium to polar substituents; it tells us nothing about the absolute value of the equilibrium constants. Thus, a large ρ-value could relate to very small equilibrium constants which change dramatically as polar substituents are introduced into a reactant, and a ρ-value close to zero could relate to very large equilibrium constants which hardly change as polar substituents are introduced.

arylsulfinate. A positive ρ value for the reactions of a series of arenediazonium cations ArN_2^+ with phenylsulfinate itself, $PhSO_2^-$, in methanol has been observed ($\rho = 3.76$).

3.5 Uses of σ-values and the Hammett equation for equilibria

The σ-value for a substituent is defined empirically by eqn 3.3 and expresses the ability of the substituent to modify the acidity of benzoic acid. This effect of the substituent reflects not only the intrinsic electronic structure and properties of the substituent, but also the mechanism by which the substituent interacts with the carboxyl group. For example, fluorine is the most electronegative element and in many molecular contexts exerts a very powerful electron-attracting effect. However, $\sigma_{(3\text{-}F)} = 0.34$ and $\sigma_{(4\text{-}F)} = 0.06$. The former reflects only a modest effect (similar to that of iodine) and the latter is surprisingly small. Whilst these parameters do not provide a molecular explanation, by quantifying an effect, they identify an issue which simplistic notions of qualitative organic chemistry cannot easily accommodate. In another case, especially with a substituent containing a less common element, its σ-value may be the most accessible experimental evidence providing an insight into the substituent's internal electronic structure and how its properties are transmitted to other parts of a molecule.

The determination of the ρ-value for a reaction series requires the measurement of equilibrium constants for, perhaps, four or five reactions under the same experimental conditions. Even this can be quite a substantial amount of laboratory work. However, from the ρ-value which is obtained and a compilation of σ-values, equilibrium constants can be easily calculated for hundreds of further members of the family, some of which may be difficult to measure experimentally (for example, if they are very large or very small).

The mechanistic importance of ρ for equilibrium constants of a reaction series springs mainly from a comparison between it and the ρ-value for the rate constants for, say, the forward direction. The former is a function of the change between initial state and final state (i.e. between reactant and product molecules), and the latter is a measure of the change between initial state and transition state (i.e. between reactant molecule(s) and activated complex). But first we need to consider the Hammett equation related to kinetics.

3.6 The Hammett equation applied to kinetics

So far, we have seen linear plots of $\log(K^X/K^H)$ for families of equilibria involving substituted aromatic compounds against σ-values of the substituents. Linear plots of $\log(k^X/k^H)$ against the same σ-values of substituents are also often obtained where k^X is the rate constant for the X-substituted member of a family of aromatic compounds, eqn 3.14, and k^H is that of the parent compound (X = H).

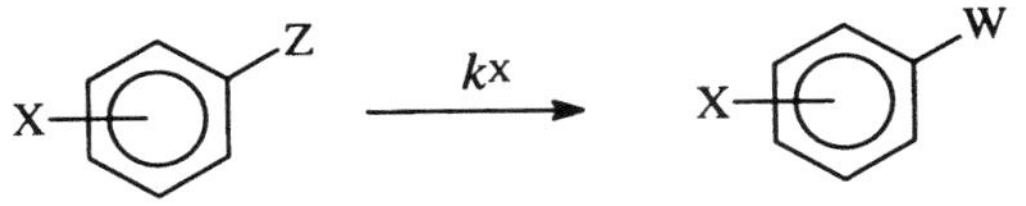

$$X\text{—}\underset{}{\bigcirc}^Z \xrightarrow{\ k_x\ } X\text{—}\underset{}{\bigcirc}^W \qquad (3.14)$$

For example, Fig. 3.6 is the Hammett plot for the alkaline hydrolysis of a series of ethyl benzoates in ethanol-water (85:15 by weight), eqn 3.15, and the gradient is 2.54.

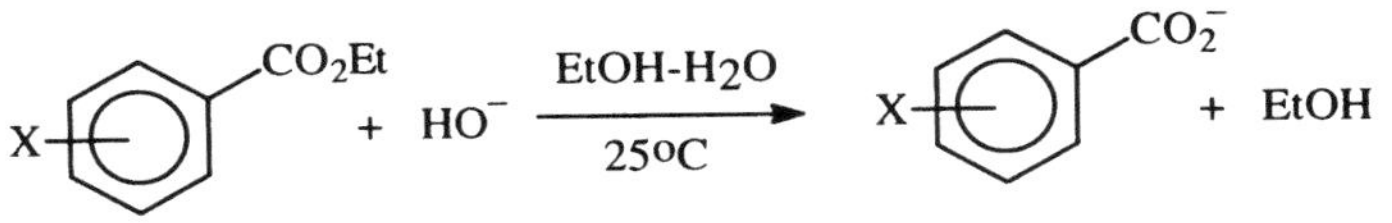

$$X\text{—}\underset{}{\bigcirc}^{CO_2Et} +\ HO^- \xrightarrow[25^\circ C]{EtOH\text{-}H_2O} X\text{—}\underset{}{\bigcirc}^{CO_2^-} +\ EtOH \qquad (3.15)$$

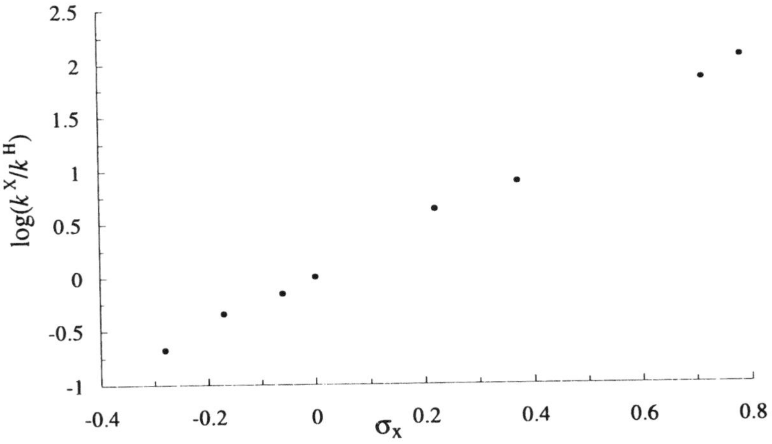

Fig. 3.6 Hammett plot for the rate constants of the reactions of eqn 3.15

Hammett rate correlations for many families of reactions have been established, and eqn 3.16 is the generalised mathematical form of such correlations

$$\log\left\{\frac{k^X}{k^H}\right\} = \rho.\sigma_x \qquad (3.16)$$

where k^X is the rate constant for the X-substituted compound,
         k^H is the rate constant for the parent member of the series (X = H),
         σ_X is the substituent parameter defined previously in eqn 3.3,
and ρ is the reaction parameter for the kinetics of the family of reactions represented here generically by eqn 3.14.

The Hammett equation for rates is not restricted to reactions of any particular rate law, and the units of the rate constants k^X and k^H cancel out in eqn 3.16.

Determination of the ρ-value for rate constants

The Hammett equation for kinetics may also be written in the 'intercept' form, i.e.

$$\log k^X = \rho.\sigma_X + \log k^o$$

in which X = H is treated just like any other substituent on the left hand side, and $\log k^o$ is the intercept in the graph of $\log k^X$ against σ_X (approximately the same value as $\log k^H$).

In principle, rate constants for perhaps four or five members of the family of compounds generalised by eqn 3.14 need to be measured under the same experimental conditions of solvent, temperature, etc. Equation 3.16 is then used to plot $\log(k^X/k^H)$, or just $\log k^X$, against the σ_X, and the gradient of the plot is the ρ-value for the reaction series.

If the rate constants change dramatically along the series (ρ is large), it may not be possible in practice to measure k^X at the same temperature for all four or five compounds. At temperatures at which the least reactive compounds proceed at a convenient rate, the rates of the most reactive analogues may be far too fast for reliable rate constant determinations. The normal procedure then is to measure rate constants over a range of temperatures and, using the Arrhenius equation, calculate the rate constant for each compound at a common temperature, usually 25°C.

The sign and magnitude of ρ for rate constants

Substituents which increase rate constants decrease the free energy of activation.

The significance of the sign and magnitude of ρ-values for rate constants is exactly analogous to that for equilibrium constants. The magnitude of the ρ-value (positive or negative) indicates the sensitivity of the rate constant to polar substituents. A family of reactions, eqn 3.14, which have the same rate constant under the same experimental conditions regardless of the nature of the substituent X in the benzene ring has a ρ-value of zero.

Positive ρ-values. A positive ρ-value for a reaction series corresponds to an increase in reactivity caused by substituents with positive σ-values (electron-withdrawing groups), or a decrease caused by substituents with negative σ-values (electron-donating groups). This requires that the free energy of activation (the free energy difference between initial state and transition state) is reduced by the introduction of electron-withdrawing groups (or increased by electron-donating groups). Either way, there must be an increase in electron density at the reaction site of the reactant bearing the substituents in the formation of the activated complex. This increase in negative charge will be dispersed by electron-withdrawing groups (but inhibited by electron-donating substituents) to an extent dependent upon the distance and relationship between the substituent and the reaction site.

$$Ar-\overset{\overset{\displaystyle O}{\|}}{C}-Cl \;+\; MeOH \;\xrightarrow[0^oC]{MeOH}\; Ar-\overset{\overset{\displaystyle O}{\|}}{C}-OMe \;+\; HCl \qquad (3.17)$$

The ρ-value is +1.42 for the methanolysis of aroyl chlorides represented in eqn 3.17, i.e. introduction of electron-withdrawing groups into the phenyl ring increases the reactivity of the acid chloride. The positive ρ-value tells us that there must be a modest increase in electron density at the carbon bonded to the arene ring in the transition state, and hence rules out a dissociative mechanism. It is compatible with an addition-elimination mechanism with rate-determining formation or decomposition of the tetrahedral intermediate.

Figure 3.7 shows representations of the uncharged tetrahedral intermediate, T, and the activated complexes for its formation and decomposition. Thus, ‡1 is the activated complex for nucleophilic attack by MeOH at ArCOCl, and ‡2 is the activated complex for departure of the chloride from T. On the grounds that chloride is the better nucleofuge, the forward reaction from T will be over a lower barrier than the back reaction, so the initial nucleophilic attack is rate determining in the overall reaction.

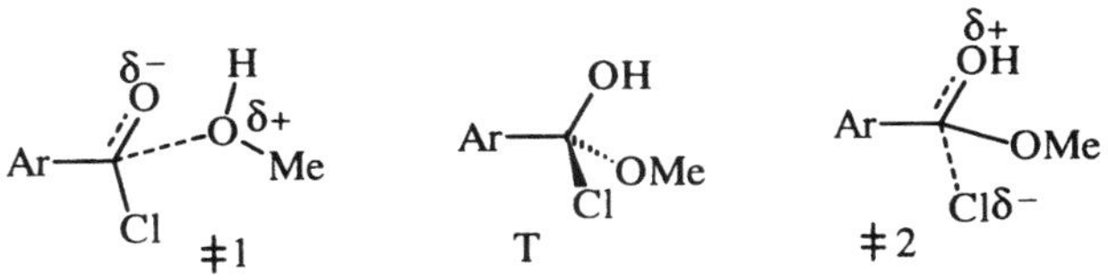

Fig. 3.7 Tetrahedral intermediate and activated complexes in the reaction of eqn 3.17

Proton transfers are involved as ‡1 becomes T in Fig. 3.7, and as ‡2 becomes product. These will be via the solvent and probably concerted with the heavy atom reorganisations rather than separate steps.

The ρ-value for the reaction of eqn 3.17 is smaller than that for the alkaline hydrolysis of ethyl benzoates (eqn 3.15) reflecting a larger build-up of negative charge in the formation of the transition state in the ester hydrolysis, as expected for the reaction involving the negatively charged nucleophile. This deduction is valid because of identical relationships between reaction site and substituents in the two families of reactions.

The ρ-value for the alkaline hydrolysis of another family of esters, $ArCH_2CO_2Et$, in aqueous ethanol at 30°C is only 0.82 compared with 2.54 for ethyl benzoates in Fig. 3.6. The lower ρ-value is because the site of the nucleophilic attack in $ArCH_2CO_2Et$ is insulated from the facilitating effect of electron-withdrawing substituents in the arene ring by the methylene group.

Comparison of the ρ-values of the hydrolyses of the two families of esters does not allow a deduction about the relative build-up of charge in the transition states of the two reactions because the relationships between reaction site and substituents are different in the two series.

Negative ρ-values. A negative ρ-value corresponds to a decrease in reactivity caused by substituents with positive σ-values (electron-withdrawing groups), or an increase in reactivity caused by electron-supplying groups. Either way, the formation of an activated complex must involve a decrease in electron density at the reaction site in the reactant bearing the substituents.

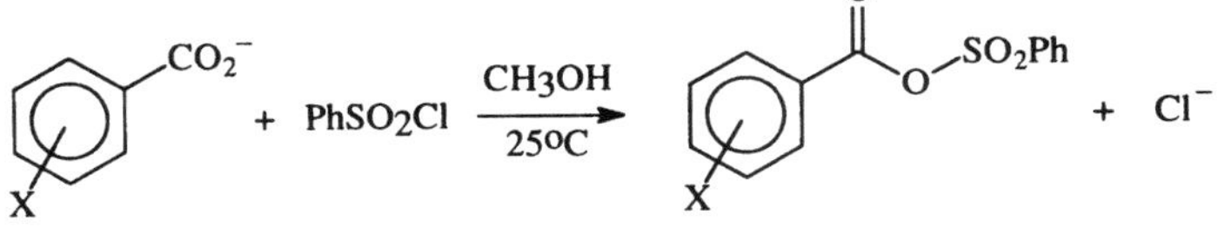

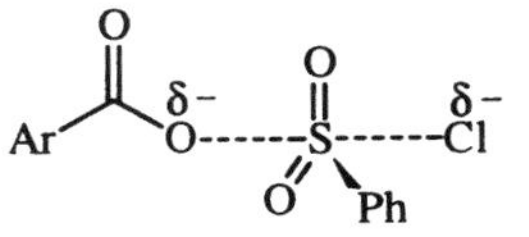

Fig. 3.8 Activated complex in the reaction of eqn 3.18

The ρ-value for the effect of substituents in the nucleophile of the substitution reaction of eqn 3.18 is -0.35. Any finite value confirms that the substitution is bimolecular, and the negative result confirms the qualitative expectation that electron-withdrawing substituents in the carboxylate make it a poorer nucleophile. Use of the Hammett equation, however, allows comparison with other reactions.

An S_N1 mechanism of the benzenesulfonyl chloride would lead to a ρ-value of zero for the nucleophiles.

For the dissociation of benzoic acids in methanol, $\rho = 1.5$. It follows that the effect of substituents upon this reaction written the other way round, i.e. as in eqn 3.19, is given by $\rho = -1.5$.

$$\text{X} \!-\! \bigcirc \!-\! \text{CO}_2^- \; + \; \text{MeO}\overset{+}{\text{H}}_2 \; \rightleftharpoons \; \text{X} \!-\! \bigcirc \!-\! \text{CO}_2\text{H} \; + \; \text{MeOH} \qquad (3.19)$$

We see, therefore, that the ρ value for the reaction of eqn 3.18 is only about one quarter of that for the equilibrium of eqn 3.19 in the same solvent. The ρ value for the former reflects the difference between arenecarboxylate and activated complex, Fig. 3.8, whereas that for the latter relates to the difference between arenecarboxylate and arenecarboxylic acid. If we may assume that the ρ value for the *overall* reaction of eqn 3.18 will be similar to the value for the overall reaction of eqn 3.19 (ca. -1.5), then the bond between the carboxylate oxygen and sulfur is only about one quarter formed in Fig. 3.8, and the carboxylate still bears a substantial negative charge.

Whether there is an increase or decrease in charge (or neither) at the sulfur in the transition state of the reaction of eqn 3.18 could be investigated by measuring the ρ value for substituents in the benzenesulfonyl chloride in reactions with a common arenecarboxylate.

3.7 Modified substituent constants in equilibrium and rate constant correlations

It follows from the definition of σ in eqn 3.3 that a simple Hammett plot is in effect a correlation between $\log K^X$ or $\log k^X$ for the family of reactions under consideration against $\log K_a{}^X$ for the family of substituted benzoic acids. Consequently, the quality of the correlation depends upon how the relationship between substituent and reaction site in the family of reactions under consideration compares with the relationship between the substituent and the carboxyl group in the family of substituted benzoic acids. If the mechanisms by which substituents interact with the reaction site in a family of reactions are just the same as those by which substituents in the benzoic acid affect its acidity, then a good correlation may be expected. On the other hand, some substituents may be able to influence reactivity in a family of reactions by a mechanism not available to the same substituents in the dissociation of substituted benzoic acids. In such reactions, the ordinary σ-values may not lead to a good correlation.

Electron-withdrawing resonance effects and σ^- parameters

A 3- or 4-nitro substituent increases the acidity of benzoic acid, and this is ascribed to the ability of the polar nitro group to disperse the negative charge of the carboxylate, an effect transmitted either through space or through the σ-bonds. A nitro substituent also enhances the acidity of phenylethanoic acid as already seen. The effect is *qualitatively* the same but smaller because of the CH_2 which intervenes between the nitrophenyl and the carboxylate.

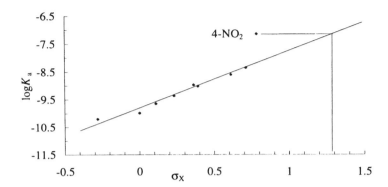

Fig. 3.9 Hammett plot for the acidity of substituted phenols in water at 25°C

Consider the acidity of a family of phenols. A normal correlation between $\log K_a$ and σ_X is obtained for 3-substituents and some 4-substituents, Fig. 3.9, and the ρ-value is quite high (2.2) because the reaction site is closer to the ring which bears the substituents than in the benzoic acid dissociation.

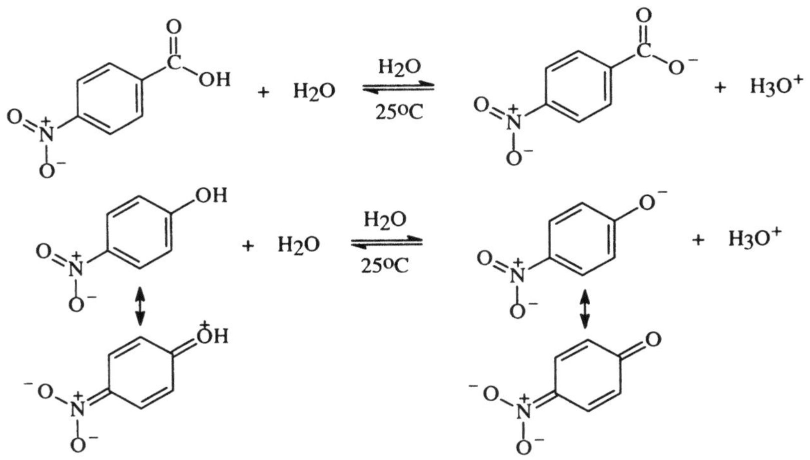

Fig. 3.10 Electron-accepting resonance effect of nitro upon the acidity of 4-nitrophenol compared with the absence of the effect upon the acidity of 4-nitrobenzoic acid

The 4-nitro group interacts by resonance with an oxygen lone-pair in the phenolate and in the phenol as shown. However, the effect will be much larger in the phenolate, so this resonance effect promotes ionization.

However, we also notice that the point for 4-nitrophenol is well above the line defined by the other compounds–it is appreciably more acidic than anticipated on the basis of the normal Hammett σ-value for the 4-nitro group (0.78). The reason is shown in Fig. 3.10. The 4-nitro group has an electron-withdrawing resonance interaction with the reaction site in the dissociation of the phenol. This interaction between substituent and reaction site is not possible in the dissociation of 4-nitrobenzoic acid, so does not contribute to

The line in the Hammett plot for substituted phenols (ρ = 2.2) is determined first using meta-substituents and para-substituents with no electron-accepting resonance effect.

the normal σ-value of 4-nitro. Enhanced acidity is observed for other phenols with substituents in the para position which are able to accept electron density by resonance, e.g. -CN, -CO$_2$R, -CHO, and -COCH$_3$.

Just as the dissociation of substituted benzoic acids in water at 25°C was used to assign σ-values for substituents without resonance effects, the dissociation of phenols in water at 25°C is used to assign new parameters to substituents with electron-attracting resonance effects. Thus, a horizontal line is drawn to the right from the point for 4-nitrophenol in Fig. 3.9 until it intersects the correlation, then a vertical line is dropped onto the σ$_X$-axis to give a new substituent parameter (σ$^-$ = 1.27) for the 4-nitro group. This parameter and corresponding values for analogous substituents are used in reactions when resonance is possible between electron-accepting substituents and a reaction site which can conjugatively supply electrons. Some σ$^-$-values are shown in Table 3.3.

p-Substituent	σ$_P$	σ$^-$
NO$_2$	0.78	1.27
CN	0.70	0.88
CH$_3$CO	0.48	0.84
CO$_2$Et	0.45	0.74
CHO	0.44	1.04
CO$_2$H	0.44	0.78
Ph	~0	0.08

Table 3.3 σ and σ$^-$ parameters of some para-substituents

The dissociation of substituted anilinium cations in water is shown in eqn 3.20. The lone pair on the nitrogen of the amino group on the product side as written may be delocalised into para-substituents such as -NO$_2$ and -CO$_2$Me.

In the equilibrium of eqn 3.20, the resonance interaction between electron-withdrawing para-substituents and the reaction site is possible only on the product side for the reaction as written. It reduces the base strengths of anilines, i.e. it enhances the acidity of the anilinium cations. The effect, therefore, is parallel with the effect of such substituents in phenols.

$$X\text{—}C_6H_4\text{—}\overset{+}{N}H_3 + H_2O \underset{25°C}{\overset{K_a^X}{\rightleftharpoons}} X\text{—}C_6H_4\text{—}\ddot{N}H_2 + H_3O^+ \quad (3.20)$$
$$\rho(\sigma^-) = 2.9$$

Consequently, in this equilibrium correlation, σ$^-$ rather than σ parameters are used as indicated; in fact, this reaction may be used to assign σ$^-$-values.

For the dissociation of phenols in ethanol, ρ(σ$^-$) ~ 2.5, so the value for the reverse reaction of a phenolate accepting a proton is ca –2.5 in ethanol. The ρ(σ$^-$)-value for the reaction of eqn 3.21 is less than half of this. So, by the argument presented on p. 38 for the S$_N$2 reaction of benzoates with benzenesulfonyl chloride, we deduce that the phenolate oxygen retains more than half its original charge in the transition state for the reaction of eqn 3.21.

$$X\text{—}C_6H_4\text{—}O^- + Et\text{-}I \xrightarrow[\text{EtOH, 43°C}]{kx} X\text{—}C_6H_4\text{—}O\text{-}Et + I^- \quad (3.21)$$
$$\rho(\sigma^-) = -0.91$$

The rate correlation for the series of eqn 3.21 also uses σ$^-$ rather than σ because, here also, there can be delocalisation of a lone pair on the oxygen at the reaction site with para substituents capable of electron-acceptance by resonance. However, in this reaction there can also be resonance in the transition state (though less than in the anionic reactant) since the oxygen retains lone pairs in the transition state. Consequently, the difference due to resonance between initial state and transition state in this reaction will be smaller than between initial and product states in the reaction of eqn 3.20, which contributes to a numerically smaller (and now negative) ρ value.

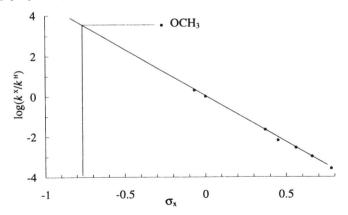

Fig. 3.11 Mechanism of nucleophilic aromatic substitution

In Fig. 3.11, we have a family of reactions with a common nucleofuge Z⁻, a common nucleophile Y⁻, a common group W which activates the reactants towards nucleophilic substitution, and substituents X which characterise the individual members of the reaction series. Monosubstituted benzene derivatives (W = X = H) do not undergo these reactions.

Nucleophilic aromatic substitution. The most important class of reactions whose mechanisms have been studied using $\rho(\sigma^-)$-correlations is nucleophilic aromatic substitution. The general mechanism is outlined in Fig. 3.11, and a particular reaction series is shown in eqn 3.22. Note the relatively high ρ value for nucleophilic attack directly at the benzene ring.

$$\rho(\sigma^-) = 3.9 \tag{3.22}$$

Electron-donating resonance effects and σ⁺ parameters

The Hammett plot for rate constants of solvolysis of X-substituted *tert*-cumyl (2-arylpropan-2-yl) chlorides in acetone-water (9:1) is shown in Fig. 3.12.

Fig. 3.12 Hammett plot for hydrolysis of *tert*-cumyl chlorides in acetone-water, 25°C

Compounds with para-substituents able to supply electron density by resonance show appreciable deviations as illustrated for the 4-methoxy compound; they are considerably more reactive than anticipated on the basis of the ordinary σ-values of their substituents. The S_N1 mechanism for this reaction is outlined in Fig. 3.13, and we see that the intermediate carbenium ion has an extra resonance form if there is a lone pair on the atom of the substituent X directly bonded to the para position of the benzene ring.

The resonance interactions shown in Fig. 3.13 are for the intermediate in the S_N1 mechanism. However, by a Hammond effect argument (see chapter 1), we deduce that the transition structure is close within the reaction coordinate to the unstable intermediate and hence an appreciable degree of resonance is involved in the transition state.

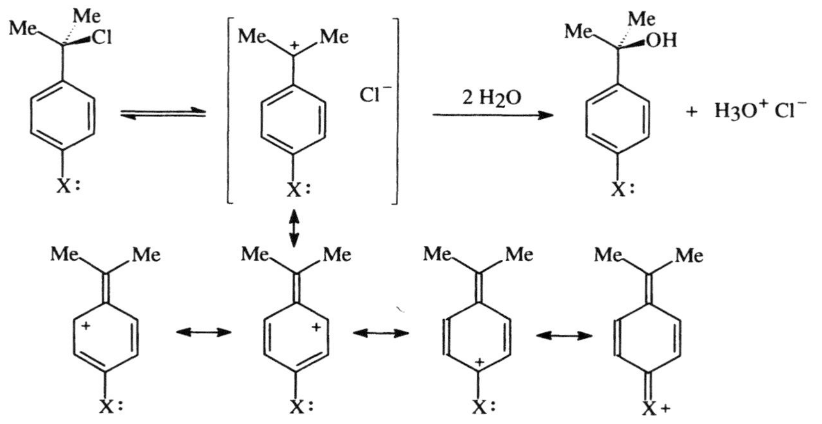

Fig. 3.13 Mechanism for hydrolysis of *tert*-cumyl chlorides in aqueous acetone, 25°C

Notice that σ⁺ parameters are assigned from a kinetics Hammett correlation whereas the σ⁻ parameters are assigned from an equilibrium.

Clearly, for reactions of this type where a para substituent has an electron-donating resonance effect which can interact with the reaction site, the ordinary σ parameter is inadequate. The strategy described above for assigning σ⁻ values is employed. A line is drawn horizontally from the point for the 4-methoxy compound in Fig. 3.12 to the correlation defined by the 3-substituted compounds and 4-substituted ones without electron-donating resonance capabilities. Then a vertical line down to the σ_X-axis allows evaluation of the new substituent parameter, σ⁺. Table 3.4 includes some σ⁺ values.

p-Substituent	σ_p	σ^+
I	0.28	0.13
Br	0.23	0.15
Cl	0.22	0.11
F	0.06	-0.07
Ph	~0	-0.21
OCH₃	-0.28	-0.78
OH	-0.38	-0.92
NH₂	-0.57	-1.3
NMe₂	-0.63	-1.7

Table 3.4 σ and σ⁺ parameters of some para-substituents

Not surprisingly considering the provenance of σ^+ parameters, many S_N1 reactions in which the nucleofuge departs from a carbon bearing a substituted phenyl group correlate with σ^+ rather than with the ordinary σ, e.g. eqn 3.23.

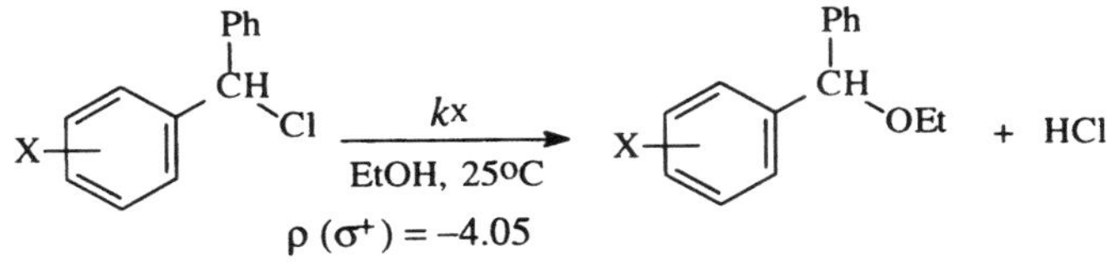

$$\rho\,(\sigma^+) = -4.05 \qquad (3.23)$$

Electrophilic aromatic substitution. One of the most important classes of reactions whose mechanisms have been studied using $\rho(\sigma^+)$-correlations is electrophilic aromatic substitution. A generic reaction and its mechanism are illustrated in Fig. 3.14, and a significant difference between this and nucleophilic aromatic substitution shown in Fig. 3.11 is immediately evident.

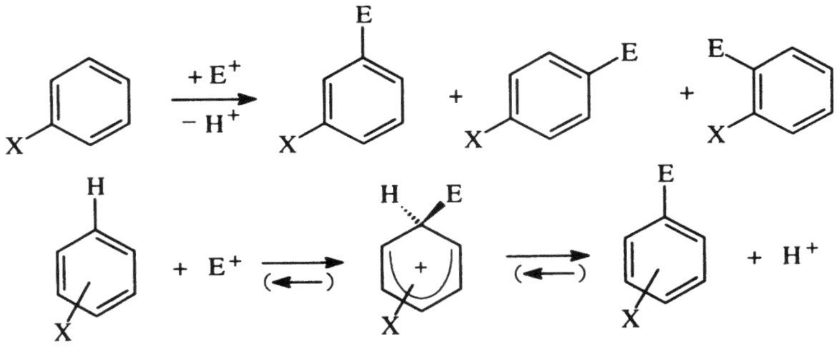

Fig. 3.14 Mechanism of electrophilic aromatic substitution

Nucleophilic substitution only takes place if there is an effective nucleofuge such as halide in the arene ring; consequently, there is no question of where the substitution will occur, and a single product is obtained. In electrophilic aromatic substitution, the proton is an effective electrofuge; consequently, in a monosubstituted reactant Ph-X, substitution may occur ortho, meta, or para to the original substituent as indicated in the upper part of Fig. 3.14. Note, however, that the three isomeric products $X\text{-}C_6H_4\text{-}E$ are formed in parallel independent reactions and do not involve a common intermediate.

The rate of the bimolecular elementary reaction between the aromatic compound and the electrophile will be second order so, if the electrophile is a stable species such as a halogen molecule, the overall reaction will be second order. However, in some reactions such as nitration in organic solvents, the generation of the electrophile itself (NO_2^+) may be the rate-determining step. Consequently, kinetic analysis of electrophilic substitution is not always straightforward. It is often investigated by product analytical competition methods and is usually discussed in terms of partial rate factors for substitution at particular positions in the aromatic ring. Thus, in the reaction with an electrophile of a mixture of known concentrations of, say, toluene or nitrobenzene and benzene, the relative rate constants at each of the three positions of the toluene or nitrobenzene compared with that of a single position in benzene can be determined from the relative yields of each of the products. For electrophilic aromatic substitution, therefore, the Hammett equation may be written as eqn 3.24, and examples are shown below.

$$\log f = \rho.\sigma^+ \tag{3.24}$$

The partial rate factor, f, for a single position in an aromatic compound ArH is given by

$$f = \frac{\text{rate constant for a single position in ArH}}{\text{rate constant for a single position in benzene}}$$

$$\rho\,(\sigma^+) = -6.2$$

$$\rho\,(\sigma^+) = -12$$

Large negative ρ-values are typical of electrophilic aromatic substitution reactions since the electrophile bonds directly to the benzene ring.

A common feature in studies of mechanisms of electrophilic substitution of aromatic compounds is the effect of substituents upon the reaction, and use of ρ-values allows the effects to be quantified. We see in the two examples above that the effect of substituents upon bromination is much greater than upon nitration under the specified conditions. We conclude that there is a greater development of positive charge in the arene ring in the transition state for bromination than for nitration.

3.8 Applications of the Hammett equation for kinetics

The reaction parameter ρ for a family of reactions may be determined from rate constants for only about four or five members of the series. From the ρ-value, rate constants may then be easily calculated for hundreds of other members of the family of compounds in the same reaction.

If ρ for the equilibrium constant of a reversible reaction is known, then ρ for the rate constant in one direction allows ρ for the other rate constant to be calculated since
$\rho(K) = \rho(k_f) - \rho(k_r)$.
And, relying upon Hammond effect arguments (see p. 11), comparison of $\rho(k_f)$ with $\rho(K)$ may provide information about the position of the transition structure within the reaction coordinate.

The ρ-value for a reaction series relates to differences between initial states and transition states, and of course we know the structures of reactants. Consequently, a Hammett investigation provides information about the structures of activated complexes. Such information is not easily available by other experimental methods.

3.9 Restrictions to the Hammett equation

The use of the simple Hammett equation is restricted to reactions of aromatic compounds, and it is observed experimentally that results for ortho-substituted members of the family of reactants (i.e. using σ_{ortho} values simply based upon the dissociation of 2-substituted benzoic acids) seldom fit onto the Hammett plot defined by results for meta and para compounds.

The effect upon the rate or equilibrium constant of a reaction of a compound of introducing a substituent will invariably be the resultant of a number of electronic and steric effects, as well as solvation effects. The very close proximity of an ortho substituent to the reaction site, especially in reactions directly at the aromatic ring, is probably why effects here, and their transmission to the reaction site, are different from effects at meta and para positions. There has been considerable progress in recent years in dissecting overall effects of substituents into constituent steric, inductive, and resonance contributions and, by such an approach, effects of ortho substituents are beginning to be understood. However, such matters are beyond the scope of the present book.

Even using only meta and para substituents, some reactions still do not give linear correlations of $\log k$ or $\log K$ against σ, σ^-, or σ^+. Occasionally, the graph indicates a trend in reactivity, but hardly qualifies as a linear plot; in some cases, no interpretable trend at all is discernible. The distinction between families of reactions which do give good Hammett plots and those which do not, however, is not a sharp one. Consequently, the quality of the Hammett correlation has to be taken into account when making mechanistic deductions.

3.10 Non-linear Hammett correlations

Occasionally, a Hammett correlation for a family of reactions appears to be a smooth curve, or shows a discontinuity at a particular value of σ, σ^-, or σ^+. The sketches in Fig. 3.15 are generic illustrations (and could have been drawn with negative gradients). The first plot shows that ρ continuously changes as σ increases, i.e. the structure of the activated complex changes along the reaction series so the mechanism is changing gradually as σ increases. The latter indicates two ρ-values according to the range of σ-values, and indicates an abrupt change in mechanism at a particular point along the series. For example, such a plot would be observed if there is a change in rate-determining step in a two-step mechanism. Addition-elimination reactions of nucleophiles with carbonyl compounds are examples, and some are considered in the next chapter.

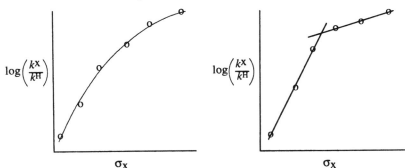

Fig. 3.15 Non-linear Hammett correlations

The difference between these two non-linear plots is exaggerated for the sake of this illustration; often it is difficult to decide whether the results are best interpreted as one or the other. In extreme cases, V- or U-shaped curves (or their inversions) are obtained, i.e. ρ may change its sign along the series.

3.11 The Yukawa-Tsuno equation

We saw above that modified σ-parameters (σ^+ or σ^-) were introduced to accommodate resonance interactions between para substituents and the reaction site. Regardless of which substituent parameter is used, the Hammett equation remains a two-parameter correlation, and the value of the substituent constant is independent of the reaction in which it is used. However, there are reactions in which neither the ordinary parameter nor the modified one leads to a satisfactory linear correlation. It appears that something between the two is required to acknowledge that resonance is not an all-or-nothing effect. However, to ascribe to each substituent a range of parameter values according to the nature of the reaction would be to remove the independence of the substituent parameter from the reaction in which it is used. A much better alternative is to extend the Hammett equation, and introduce a third parameter, r^+ or r^- according to whether the substituent is π-electron-supplying or -accepting. This new reaction parameter is a measure of the extent of resonance between para substituents and reaction site, i.e. the resonance demand of the reaction site. It is incorporated in the Yukawa-Tsuno equation for rate constants given in eqn 3.25; there is also a version with r^-, and two corresponding equations for equilibrium constants.

$$\log\left\{\frac{k^X}{k^H}\right\} = \rho\{\sigma_X + r^+(\sigma_X^+ - \sigma_X)\} \tag{3.25}$$

For a family of reactions in which there can be no resonance interaction at all between substituents and the reaction site, then $r = 0$ and equation 3.25 collapses down to the normal Hammett equation. If $r^+ = 1$, eqn 3.25 may be simplified to give eqn 3.26 which corresponds to the Hammett equation using the σ^+ parameter.

$$\log\left\{\frac{k^X}{k^H}\right\} = \rho.\sigma_X^+ \tag{3.26}$$

It follows that the hydrolysis of *tert*-cumyl chlorides in Fig. 3.13 not only defines σ^+-values for substituents, but is also the standard reaction for which $r^+ = 1$. There are reactions for which $0 < r^+ < 1$, e.g. eqn 3.27,

$$C_6H_5X \ + \ HNO_3 \ \xrightarrow[\text{25}^\circ\text{C}]{CH_3CO_2H} \ XC_6H_4NO_2 \ + \ H_2O \tag{3.27}$$

$$\rho = -6.38, \quad r^+ = 0.90$$

and also ones for which $r^+ > 1$, for example, eqn 3.28.

$$XC_6H_4CPh_2Cl \ + \ CH_3OH \ \xrightarrow[\text{25}^\circ\text{C}]{CH_3OH} \ XC_6H_4CPh_2OCH_3 \ + \ HCl \tag{3.28}$$

$$\rho = -4.02, \quad r^+ = 1.23$$

The transition states in the family of reactions of eqn 3.27 involve a lower degree of resonance than do those in the hydrolysis of *tert*-cumyl chlorides, whereas the transition states in the reactions of eqn 3.28 involve more.

$$\log\left\{\frac{k^X}{k^H}\right\} = \rho.\sigma_X^-\qquad(3.29)$$

Correspondingly, if $r^- = 1$, the Yukawa-Tsuno equation gives eqn 3.29 which is the Hammett equation using the modified σ^- parameter. So here we see that the dissociation of phenols in water at 25°C is the standard reaction with $r^- = 1$ as well as the reaction which defines the σ^- parameter. The acid dissociation in eqn 3.30 has $r^- = 0.54$, i.e. the resonance demand in this equilibrium is only about half that in the acid dissociation of phenols. In this reaction, the strong electron-withdrawing effect of the sulfonyl group stabilises the anion and thereby facilitates the dissociation. This effect is not present in the dissociation of phenols (Fig. 3.10) which makes a greater demand, therefore, on the resonance effect of substituents X.

$$\rho = 1.88, \quad r^- = 0.54$$

Problems

3.1 For the dissociation of substituted benzoic acids in dimethyl sulfoxide at 25°C, $\rho = 2.48$ and $pK_a = 11.0$ for the parent compound. Calculate pK_a values of the 3-NO_2 and 4-OMe analogues in DMSO (σ-values on p. 32).

3.2 Determine the reaction parameter, ρ, for the reaction shown below from the data provided (σ-values on p. 32), and propose a mechanism.

Substituent, X	$k/dm^3\ mol^{-1}\ s^{-1}$	Substituent, X	$k/dm^3\ mol^{-1}\ s^{-1}$
4-MeO	3.01×10^{-2}	4-Br	6.12×10^{-4}
4-Me	1.12×10^{-2}	3-Cl	2.66×10^{-4}
3-Me	6.00×10^{-3}	3-NO_2	2.86×10^{-5}
H	3.44×10^{-3}		

(a) Comment upon the probable effect of X = 4-NO_2 upon this reaction.

(b) For the reaction of the parent compound (X = H), $\Delta H^{\ddagger} = 47$ kJ mol^{-1} and $\Delta S^{\ddagger} = -146$ J K^{-1} mol^{-1}; comment on these results in the light of your mechanism.

3.3 Rate constants at 20°C and substituent parameters are given for the nucleophilic addition reaction in methanol below. Construct appropriate

Hammett plots and evaluate ρ for k_1 and k_{-1}. Hence or otherwise, determine ρ for the equilibrium constant of the reaction ($K = k_1/k_{-1}$).

Substituent X	σ	σ^-	$k_1/\text{dm}^{-3}\,\text{mol}^{-1}\,\text{s}^{-1}$	k_{-1}/s^{-1}
CF_3SO_2	0.76	1.36	141	1.17×10^{-4}
NO_2	0.78	1.27	11.8	6.05×10^{-4}
CN	0.70	1.00	2.82	1.68×10^{-2}
CH_3SO_2	0.64	0.98	1.75	1.68×10^{-2}
CF_3	0.54	-	0.40	8.0×10^{-2}
Cl	0.22	-	1.2×10^{-2}	5
F	0.06	-	2.5×10^{-3}	30
H	0	0	1.5×10^{-3}	20

3.4 From the following rate results for the solvolysis of substituted benzyl azoxyarenesulfonates (I) at 25°C in aqueous trifluoroethanol, calculate reaction parameters for the effects of substituents X and Y using the appropriate substituent constants.

X in I, Y = Me	$10^5 k/\text{s}^{-1}$
3-Cl	0.031
4-Cl	0.23
H	0.47
3-Me	0.76
4-Me	6.0
4-OMe	170

Y in I, X = 4-Me	$10^5 k/\text{s}^{-1}$
OMe	4.24
Me	6.0
Br	19.1
CN	46.5

For the S_N1 solvolysis of substituted benzyl tosylates (II), $\rho(\sigma_X^+) = -5.6$, and for 2-adamantyl arenesulfonates (III) $\rho(\sigma_Y) = 1.9$; hence, what may be deduced from the ρ-values for solvolysis of I regarding its mechanism?

4 Catalysis of organic reactions in solution by small molecules and ions

4.1 Introduction

Catalysis is the enhancement of the rate of a reaction by a compound (the catalyst) not generally present in the chemical equation which describes the reaction. Normally, a catalyst remains unchanged by the chemical reaction it catalyses. It brings about the rate enhancement by providing a reaction pathway additional to the one which occurs in its absence. This additional pathway will have its own rate law, and the total rate of reaction in the presence of the catalyst is the sum of the catalysed and uncatalysed pathways as illustrated for a generic second-order reaction in Fig. 4.1.

The catalyst provides an additional pathway for the reaction, i.e. the products are not changed and neither is the equilibrium constant if the reaction is reversible.

$$A + B \xrightarrow{\;k\;} C + D$$
rate of uncatalysed reaction pathway $= k[A][B]$

$$A + B \xrightarrow[X]{\;k_X\;} C + D$$
rate of reaction pathway catalysed by $X = k_X[X][A][B]$

Fig. 4.1 Uncatalysed and catalysed channels for a second-order reaction

The catalysed and uncatalysed reactions are often of the same kinetic order in reactants, and the catalysed reaction is usually first order in catalyst.

For the reactions in Fig. 4.1,

$$\text{total rate of reaction} = k[A][B] + k_X[X][A][B]$$
$$= (k + k_X[X])[A][B] .$$

Since $[X]$ remains constant during the reaction, this rate law may be written

$$\text{total rate} = k_{obs}[A][B]$$

where
$$k_{obs} = k + k_X[X] . \tag{4.1}$$

In this example, k_{obs} is the observed *pseudo* second-order rate constant.

The uncatalysed reaction may be so slow that it is undetectable ($k \sim 0$), in which case the total reaction in the presence of the catalyst is effectively just the catalysed reaction, $k_{obs} = k_X[X]$. This is very commonly the case for biological reactions involving enzymes as catalysts.

In Fig. 4.1, k_X is the catalytic constant for compound X and its effectiveness as a catalyst for the particular reaction is indicated by the ratio k_X/k. A catalysed reaction is invariably catalysed by a range of catalysts and each has its own catalytic constant; like all rate constants, each will depend upon the experimental conditions, e.g. temperature and solvent. The usual procedure for measuring a catalytic constant for a reaction represented by Fig. 4.1 is based upon eqn 4.1. Values of k_{obs} are measured by the normal methods of kinetics for different (but constant) concentrations of X, then a graph of k_{obs} against [X] gives k_X as the gradient and the intercept at [X] = 0 gives k, the rate constant for the uncatalysed reaction.

Note that in Fig. 4.1, the catalysed reaction is first order in catalyst, and k_X is a third-order rate constant. If an example which is first order when uncatalysed had been used, the catalysed reaction would have been second order overall, but the method for determining k_X is exactly the same. The procedural details for measuring a catalytic constant are different if the catalysed reaction is second order in the catalyst, but not difficult to work out.

Quite often, as we shall see, a particular compound may be an effective catalyst for different types of reactions, but usually its mode of action is similar in all cases. We shall now look at the relatively small number of types of catalytic mechanisms.

4.2 Electrophile catalysis

The hydrolysis of methyl esters of sterically congested carboxylic acids is very slow even under either acidic or basic conditions. This is because nucleophilic addition of water or hydroxide to the carbonyl group of such esters is sterically hindered. However, the compounds may be cleaved in a different type of reaction using lithium iodide in pyridine. In this reaction, iodide acts as a nucleophile and the carboxylate is the nucleofuge, i.e. it is an S_N2 displacement at the methyl group, but sodium and potassium iodide are not effective.

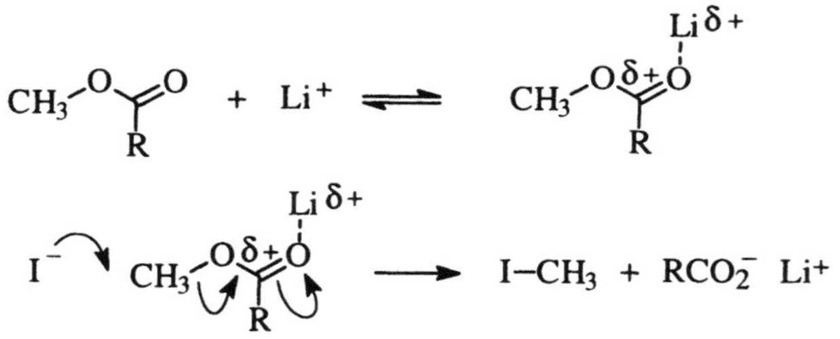

Fig. 4.2 Mechanism for the cleavage of methyl esters of sterically hindered carboxylic acids by lithium iodide in pyridine

A hard Lewis acid cation such as Li^+ is required which is complexed by an oxygen of the ester in a pre-equilibrium and, thereby, converts the carboxylate into a better nucleofuge, as indicated in Fig. 4.2 where R is a bulky alkyl residue. Catalysis by Li^+ does not occur if the reaction is carried out in hydroxylic solvents which (unlike pyridine) complex the lithium cation, i.e. they compete too effectively with the ester for the lithium cation.

The electrophile catalysis mechanism of Fig. 4.2 may be generalised as shown in Fig. 4.3. The substrate in the uncatalysed reaction (A-B) is made more electrophilic, i.e. more reactive towards attack by the nucleophilic reagent (Nu⁻), in the catalytic mechanism by complexation with the catalyst (E⁺) in a pre-equilibrium. Figure 4.3 illustrates another feature of catalysis in general, not just electrophile catalysis: although an uncatalysed reaction may be concerted, i.e. single step, catalytic mechanisms are necessarily multi-step proceeding through intermediates whose lifetimes vary widely according to the particular reaction.

$$A\text{-}\ddot{B} \ +\ E^+ \ \rightleftharpoons \ A\text{-}\overset{+}{B}\text{-}E$$

$$Nu^- \ +\ A\text{-}\overset{+}{B}\text{-}E \ \longrightarrow \ Nu\text{-}A \ +\ \ddot{B}\text{-}E$$

$$\ddot{B}\text{-}E \ \rightleftharpoons \ B^- \ +\ E^+$$

$$\text{Overall,} \quad Nu^- \ +\ A\text{-}\ddot{B} \ \xoverset{E^+}{\longrightarrow} \ Nu\text{-}A \ +\ B^-$$

Fig. 4.3 General mechanism for electrophile-catalysed nucleophilic substitution

Reactions other than substitution may be catalysed by electrophiles, e.g. nucleophilic addition reactions, and uncharged Lewis acids such as BF_3 are effective electrophile catalysts for some reactions under non-aqueous conditions. In aqueous solution, a number of metal ions act as electrophile catalysts including Zn^{2+}, Hg^{2+}, and Cu^{2+}.

In general, electrophile catalysis renders substrates more susceptible to attack by nucleophiles.

4.3 Specific acid catalysis

When electrophile catalysis involves proton transfer, i.e. the catalyst is a Brønsted rather than a Lewis acid, it is called *acid catalysis* and is very common for a wide range of reactions especially in aqueous solution. The phenomenon is further subdivided into *specific* and *general* acid catalysis according to an experimental criterion but, as we shall see, there is a fundamental mechanistic basis to this distinction. First, specific acid catalysis will be introduced through a familiar reaction, then this will be used to demonstrate how the phenomenon is investigated and how it is distinguished experimentally from general acid catalysis.

Hydrolysis of simple carboxylic esters

The hydrolysis of a simple ester such as ethyl ethanoate is reversible, and the equilibrium constant at normal temperatures is such that the reaction may be driven in either direction by an appropriate choice of experimental conditions. A low concentration of ester in water will give a very high conversion to ethanol and ethanoic acid at equilibrium, and a low concentration of ethanoic acid in ethanol will give a good conversion to ester at equilibrium. However,

The rate of the reverse of the reaction in eqn 4.2, i.e. of dilute carboxylic acid with dilute alcohol, is negligible so the observed reaction starting from the dilute ester in water is essentially just the forward reaction.

the reaction in either direction is very slow at normal temperatures in the absence of a catalyst.

Experimental evidence. Equation 4.2 represents the hydrolysis of a low concentration of the ester, E, in water with catalysis by hydronium ion.

$$CH_3CO_2Et \ (E) + \ H_2O \ \underset{(\longleftarrow)}{\xrightarrow{H_2O, H_3O^+}} \ CH_3CO_2H \ + \ EtOH \qquad (4.2)$$

The rate of this reaction in the absence of added H_3O^+ ($pH = 7$) is very slow and given by the first-order rate law

$$- d[E]/dt = k_o[E]$$

where $k_o =$ the first-order rate constant for the solvent-induced reaction.

Under acidic conditions ($pH < 7$), an additional second-order reaction is observed, and the total rate law is given by

$$- d[E]/dt = k_o[E] \ + \ k_H[H_3O^+][E]$$

or $$- d[E]/dt = (k_o \ + \ k_H[H_3O^+]) \, [E]$$

where $k_H =$ the second-order rate constant (the catalytic constant) for the reaction catalysed by H_3O^+.

During a single reaction, the concentration of hydronium ions remains constant; consequently, this equation may be written

k_{obs} is easily measured in principle by monitoring [E] with time. The gradient of the linear plot of $\ln[E]_t$ against t is $-k_{obs}$.

$$- d[E]/dt = k_{obs}[E]$$

where $$k_{obs} = k_o \ + \ k_H[H_3O^+] \qquad (4.3)$$

The reaction is now seen to have a *pseudo* first-order rate law, and k_{obs} is the observed experimental *pseudo* first-order rate constant.

Strictly, the concentration of H_3O^+ is not equal to 0 at pH = 7, but it is negligibly small compared with values in even quite dilute solutions of mineral acids.

After measurements of k_{obs} at different known constant concentrations of hydronium ion, the gradient of a graph of k_{obs} against $[H_3O^+]$ gives k_H and the intercept gives k_o, as illustrated in Fig. 4.4. This method allows determination of k_o even when it is too small to be measured directly at $pH = 7$.

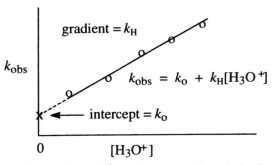

Fig. 4.4 Plot of k_{obs} against $[H_3O^+]$ for the catalysed hydrolysis of a simple ester

For the catalysed hydrolysis of a simple ester, the source of the hydronium ions does not matter. It could be from the complete dissociation of a dilute strong mineral acid such as hydrochloric acid of known concentration, or it could be from an acidic buffer solution of uncertain composition but of known pH (e.g., by measurement using a pH meter) from which the concentration of hydronium ions may be calculated.

We know from the above that the low concentration of hydronium ions from the incomplete dissociation of a weak acid will catalyse the reaction according to eqn 4.3, but what about the effect of the undissociated acid? The dissociation of a weak acid, AH, in water is shown in Fig. 4.5, and includes the relationship between the pH of the solution, the pK_a of AH, and the ratio of undissociated acid to its anion, [AH]/[A⁻] = r, the so-called *buffer ratio*. In particular, we see that it is possible to vary the *concentration* of AH at constant pH, i.e. constant [H₃O⁺], as long as the buffer *ratio* is kept constant.

$$AH + H_2O \xrightleftharpoons{H_2O} H_3O^+ + A^-$$

$$K_a = \frac{[H_3O^+][A^-]}{[AH]} \qquad [H_3O^+] = K_a \cdot \frac{[AH]}{[A^-]}$$

$$\log[H_3O^+] = \log K_a + \log\left(\frac{[AH]}{[A^-]}\right)$$

$$pH = pK_a - \log r$$

$$\text{where } r = \frac{[AH]}{[A^-]}, \text{ the buffer ratio}$$

Fig. 4.5 Relationship between pH, pK_a, and r for a weak acid in aqueous solution

In a solution of a weak acid AH, a salt of the acid (extra A⁻) may be added, or a fully dissociated strong acid (extra H₃O⁺), or even both. In all cases, the expression for the dissociation constant for AH must include total concentrations of all species at equilibrium, regardless of their origin.

It is, therefore, relatively straightforward to test experimentally whether the undissociated weak acid, AH, catalyses a reaction such as the hydrolysis of an ester. A series of k_{obs} measurements are made at different buffer concentrations, i.e. the concentrations of AH are different, but at a constant buffer ratio so the pH and hence [H₃O⁺] are kept constant. If k_{obs} does not increase as the buffer concentration is increased at constant pH, then the weak acid does not catalyse the reaction.

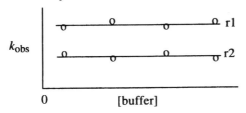

Fig. 4.6 Graphs of k_{obs} against [buffer] for two values of r for specific acid catalysis

In Fig. 4.6, k_{obs} at r1 is greater than k_{obs} at r2, i.e. the pH is lower in the solution of r1 than that of r2 (buffer ratio r1 > buffer ratio r2).

Results for the hydrolysis ethyl ethanoate at two buffer ratios, r1 and r2, i.e. two pH values, are sketched in Fig. 4.6. The reaction is catalysed by H_3O^+ but not by an undissociated weak acid, AH.

Reaction mechanism. A catalytic mechanism is required which accommodates the kinetics results presented above as well as other evidence, e.g. isotopic labelling studies. The A2 mechanism shown in Fig. 4.7 for any simple ester RCO_2R' involves a pre-equilibrium protonation of the ester by hydronium ion. The rate-limiting step is either the subsequent nucleophilic attack by water or departure of the alcohol molecule from the protonated tetrahedral intermediate; either way, it does not involve proton transfer.

The source of the hydronium ions in the mechanism of Fig. 4.7 is irrelevant; it could be a fully dissociated strong acid such as hydrochloric, or it could be the low extent of dissociation of a weak acid.

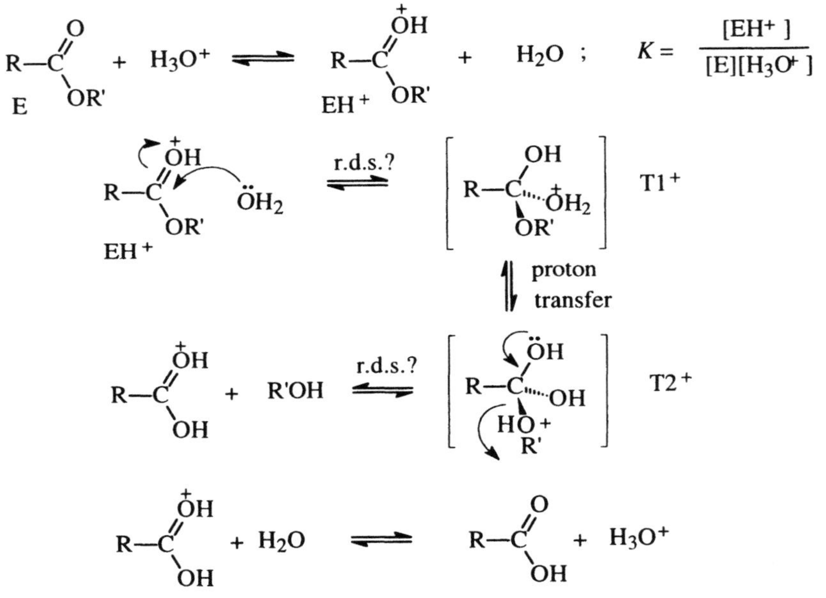

Fig. 4.7 Mechanism for the specific acid catalysed hydrolysis of a simple ester

It is easily shown that, for any reaction, not just ester hydrolysis, a mechanism involving pre-equilibrium protonation of the substrate followed by a rate-limiting step which does not involve proton transfer will lead inevitably to a specific acid catalysis rate law. This is true even if the original proton donor is an undissociated weak acid.

By analogy, we may identify the very slow uncatalysed hydrolysis with a mechanism involving nucleophilic attack by water at the carbonyl of an unprotonated ester molecule.

According to this mechanism, the rate-limiting step is either the nucleophilic attack of water at the carbonyl of the protonated ester to give the positively charged tetrahedral intermediate T1$^+$, or the departure of the alcohol molecule as nucleofuge from the isomeric charged tetrahedral intermediate T2$^+$ to leave the protonated carboxylic acid molecule. In both cases, the composition of the activated complex (though not its structure) is the same and comprises the elements of an ester molecule, a proton, and a water molecule. Consequently, the rate law of the catalytic reaction channel will be the same whichever of these two steps is rate limiting and, since water is the solvent and hence at constant concentration, the rate law is

$$\text{rate} = k[\text{EH}^+]$$

or

$$\text{rate} = kK[\text{E}][H_3O^+] .$$

The two constants may be combined to give

$$\text{rate} = k'[E][H_3O^+]$$

When this rate law derived from the mechanism in Fig. 4.7 is compared with the catalytic term in the experimental rate law on p. 52, we see they are the same with $k' = k_H$. The catalytic mechanism shown in Fig. 4.7 as chemical equations is represented as a free energy profile in Fig. 4.8.

Initial and final proton transfer equilibria shown in Fig. 4.7 are omitted from Fig. 4.8.

The uncharged tetrahedral intermediate T

$$R-\overset{\overset{\displaystyle OH}{|}}{\underset{\underset{\displaystyle OH}{|}}{C}}\cdots OR'$$

can form by deprotonation as illustrated below.

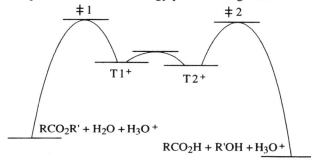

Fig. 4.8 Free energy profile for the specific acid catalysed hydrolysis of a simple ester (the A2 mechanism of Fig. 4.7)

Comparisons of the rate of hydrolysis with the rate of exchange of isotopically labelled oxygen between ester and the aqueous solvent have established that transition states ‡1 and ‡2 are of comparable height for simple esters. For other esters, either one or the other can be rate determining.

4.4 General acid catalysis

In the experimental test for whether a reaction is specific acid catalysed, k_{obs} is measured at different buffer concentrations, i.e. different concentrations of a weak acid AH, but at a constant buffer ratio, i.e. constant $[H_3O^+]$. If the results are as shown previously in Fig. 4.6, k_{obs} is independent of [AH] which, therefore, is not a catalyst; the reaction is catalysed only by H_3O^+.

In Fig. 4.9, k_{obs} increases linearly with the concentration of AH at two different constant hydronium ion concentrations. Such a reaction is catalysed by AH (as well as by H_3O^+) and is a *general* acid catalysed reaction.

$$\text{Substrate (S)} \xrightarrow[k_{obs}]{AH,\ H_3O^+,\ H_2O} \text{Products} \qquad (4.4)$$

Equation 4.4 represents such a reaction which is first order in substrate, S. If there is also an appreciable water-induced reaction, the complete rate law in the presence of hydronium ions and an undissociated weak acid AH is

$$\text{rate} = k_o[S] + k_H[S][H_3O^+] + k_{AH}[S][AH]$$

$$= (k_o + k_H[H_3O^+] + k_{AH}[AH])\,[S]$$

where k_{AH} = the catalytic constant for reaction catalysed by AH

and other terms have their previous meanings.

In Fig. 4.6, the x-axis was labelled [buffer], and in Fig. 4.9 it is [AH]. The buffer is made up from AH and its conjugate base A⁻, e.g. as the sodium salt. Consequently, the x-axis may be labelled according to the context. In another context, for example, it may be appropriate to label it [A⁻].

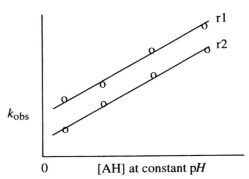

Fig. 4.9 Graphs of k_{obs} against weak acid concentration at two buffer ratios for a general acid catalysed reaction

In a single experiment, $[H_3O^+]$ and $[AH]$ are kept constant; consequently, the above rate law may be written as

$$\text{rate} = k_{obs}[S]$$

where $k_{obs} = k_o + k_H[H_3O^+] + k_{AH}[AH]$.   (4.5)

The *pseudo* first-order rate constant k_{obs} is readily determined in the usual way, i.e. from the gradient of a plot of $\ln[S]_t$ against time.

Experimental characterisation of a general acid catalysed reaction

We have seen above how to establish whether a reaction is specific or general acid catalysed. It remains to see how k_o, k_H, and k_{AH} for a general acid catalysed reaction are determined.

If the reaction is investigated in the absence of any weak acids, i.e. as though it were specific acid catalysed, $[AH] = 0$ and eqn 4.5 becomes the same as eqn 4.3. Consequently, k_o and k_H for a general acid catalysed reaction can be determined from a set of experiments in the absence of AH leading to a graph such as that in Fig. 4.4.

Next, the reaction is investigated using buffer solutions containing the weak acid AH and a salt of its conjugate base A⁻, and the results are plotted as in Fig. 4.9. For one set of experiments at a particular buffer ratio, $[H_3O^+]$ is constant. Consequently, from eqn 4.5, the gradient of the plot of k_{obs} against $[AH]$ is k_{AH} (the intercept of this plot is $k_o + k_H[H_3O^+]$).

General acid catalysis and reaction mechanisms

In another context, pre-equilibrium protonation of a substrate followed by rate-limiting deprotonation also leads to a general acid catalysis rate law (see p. 78, problem 4.3).

Establishing experimentally that a particular reaction has a general acid catalysis rate law (eqn 4.5) does not lead directly to a single mechanism. However, any mechanism which involves rate-limiting proton transfer from the catalyst to the substrate will necessarily lead to a general acid catalysis rate law. To demonstrate this, we shall consider the acid-catalysed hydrolysis of vinyl ethers shown generically in eqn 4.6 for compound E.

$$CH_2=CH\text{-}OR \text{ (E)} + H_2O \xrightarrow[k_{obs}]{AH,\ H_3O^+,\ H_2O} CH_3\text{-}CH=O + ROH \quad (4.6)$$

These compounds are stable in non-acidic aqueous conditions (so there is no water-induced reaction, i.e. $k_o = 0$) but, upon acidification, they give alcohols and carbonyl compounds. Experimentally, the reactions are *pseudo* first order in vinyl ether and general acid catalysed with $k_o = 0$,

i.e. $\qquad\qquad -d[E]/dt = k_{obs}[E]$

where $\qquad\qquad k_{obs} = k_H[H_3O^+] + k_{AH}[AH]$.

We shall consider the mechanistic proposal that the vinyl ether reacts via rate-limiting protonation to give a reactive intermediate, and that H_3O^+ *and* AH contribute parallel routes.

The first two steps in Fig. 4.10 are parallel proton transfers from hydronium ion and the general acid to the substrate to give the same reactive intermediate, an oxacarbenium ion. Each step has its own elementary rate constant. Whilst in principle these initial proton transfers are reversible, the back reactions cannot compete with the common forward reaction–the capture of the oxacarbenium ion by water to give a tetrahedral intermediate.

The forward reaction of the reactive intermediate involves nucleophilic attack by the solvent water at a cation reminiscent of a protonated aldehyde or ketone; the back reaction would involve breaking a carbon–hydrogen bond.

The loss of alcohol from the protonated tetrahedral intermediate is also reversible in principle. In practice, the forward reaction of the protonated aldehyde with water is much faster than the reverse second-order reaction with the very low concentration of alcohol.

The isomeric, positively charged, tetrahedral intermediates in this mechanism can be deprotonated to give the same hemiacetal.

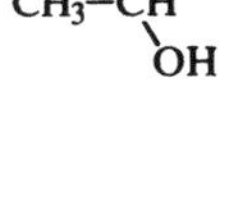

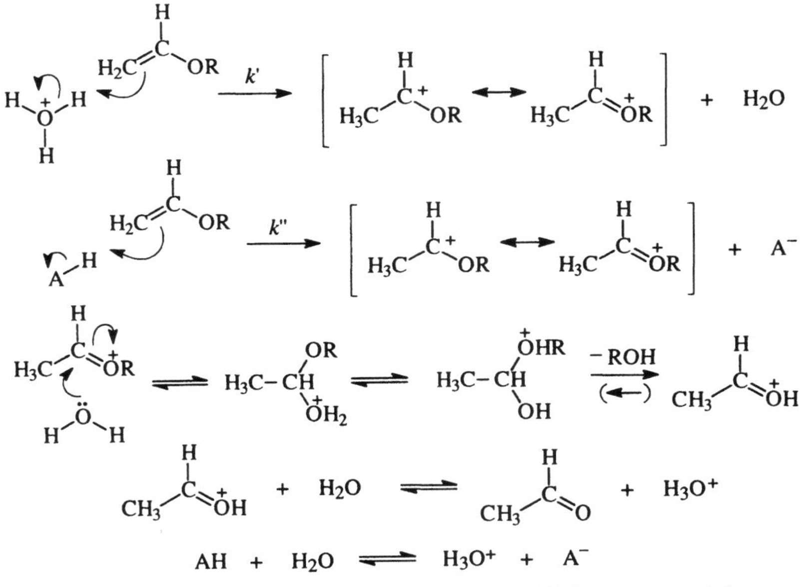

Fig. 4.10 Hydrolysis of vinyl ethers catalysed by acids in aqueous solution

According to this mechanism, the rate of the overall reaction will be the sum of the rates of the parallel reactions which generate the common reactive intermediate, i.e. the first two equations of Fig. 4.10:

$$-d[E]/dt = k'[E][H_3O^+] + k''[E][AH]$$

$$= (k'[H_3O^+] + k''[AH]) [E].$$

Thus, the reaction is predicted to be *pseudo* first order in the substrate E at constant concentrations of hydronium ion and undissociated acid, AH, as is found experimentally, and the elementary rate constants of the mechanism, k' and k'', are identical with the experimental parameters k_H and k_{AH}. If further acids A'H, A''H, etc., are present, each contributes an additional parallel reaction channel, and the experimental rate law becomes

It is assumed in this mechanism that water is not strong enough a Brønsted acid to be able to protonate vinyl ethers at a measurable rate; consequently, no solvent-induced reaction channel is predicted.

$$k_{obs} = k_H[H_3O^+] + k_{AH}[AH] + k_{A'H}[A'H] + k_{A''H}[A''H], \text{ etc.}$$

The generalised version of eqn 4.5, the general acid catalysis rate law, therefore, is eqn 4.7.

$$k_{obs} = k_o + k_H[H_3O^+] + \Sigma k_{AH}[AH] \qquad (4.7)$$

4.5 Nucleophile catalysis

The hydrolysis of a simple primary alkyl chloride is extremely slow, eqn 4.8. Water is not good enough a nucleophile to displace chloride easily in an S_N2 mechanism, and simple primary carbenium ions are too unstable so there is no viable S_N1 mechanism.

$$RCH_2\text{-}Cl + 2H_2O \xrightarrow[\text{slow}]{H_2O} RCH_2\text{-}OH + H_3O^+ + Cl^- \qquad (4.8)$$

If an ionic iodide is added to the solution, however, the primary alcohol is readily produced and iodide is present at the end just as at the beginning, i.e. I^- catalyses the hydrolysis and a mechanism is shown in Fig. 4.11.

$$RCH_2\text{-}Cl + I^- \longrightarrow RCH_2\text{-}I + Cl^-$$

$$RCH_2\text{-}I + 2H_2O \longrightarrow RCH_2\text{-}OH + H_3O^+ + I^-$$

Fig. 4.11 Mechanism for iodide catalysis of hydrolysis of a simple primary alkyl chloride

Iodide is a much better nucleophile than water, so is able to displace chloride in aqueous solution in an S_N2 reaction; then iodide, being also a good nucleofuge, is displaced in a second S_N2 reaction by water. The algebraic sum of the two steps of Fig. 4.11 is the reaction of eqn 4.8 which does not occur in the absence of the catalyst. Since the catalyst in the first step of this mechanism acts as a nucleophile, the overall mechanism is referred to as nucleophile catalysis.

The above is typical of many reactions catalysed by nucleophiles, and we may generalise as follows.

i)   The catalyst must be more nucleophilic than the nucleophile in the uncatalysed reaction.

ii) The catalyst must become a better nucleofuge in the intermediate than the original one in the reactant.

iii) The intermediate must be less stable under the reaction conditions than the product.

If (i) were not so, there would be no catalysed reaction. If (i) is true but not (ii) and (iii), a reaction will occur but not catalysis of the original reaction.

4.6 Specific base catalysis

Some reactions of organic compounds in aqueous solution are catalysed by hydroxide, the conjugate base of the solvent, but unaffected by other bases, e.g. an amine or a carboxylate salt, which may be present. Regardless of mechanistic detail, this is termed *specific* base catalysis.

In a buffered solution, a specific base catalysed reaction is dependent upon the buffer ratio, i.e. the pH, but independent of the buffer concentration. This leads to a convenient experimental procedure for establishing that a reaction is specific base catalysed. Consider the reaction of eqn 4.9 which is first order in the concentration of the substrate, S. In the presence of general base, B, and hydroxide, there are just uncatalysed (solvent induced) and hydroxide catalysed reaction channels as shown by the rate law of eqn 4.10.

$$\text{Substrate (S)} \xrightarrow[\text{B}]{\text{H}_2\text{O, OH}^-} \text{Product} \tag{4.9}$$

$$-d[\text{S}]/dt = k_o[\text{S}] + k_{\text{OH}}[\text{OH}^-][\text{S}] \tag{4.10}$$

$$= k_{\text{obs}}[\text{S}] \qquad \text{at constant p}H$$

where $\quad k_{\text{obs}} = k_o + k_{\text{OH}}[\text{OH}^-] \tag{4.11}$

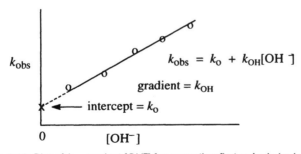

$$k_{\text{obs}} = k_o + k_{\text{OH}}[\text{OH}^-]$$
$$\text{gradient} = k_{\text{OH}}$$
$$\text{intercept} = k_o$$

Fig. 4.12 Plot of k_{obs} against [OH$^-$] for a reaction first order in hydroxide

In order to establish that a reaction is specific base catalysed, we proceed as follows. First, the *pseudo* first-order rate constant, k_{obs}, is measured in the usual manner at different constant concentrations of hydroxide, e.g. by using sodium hydroxide solutions, but with no other base present. Equation 4.11 indicates that k_{OH} and k_o are obtained from the slope and intercept of the graph of k_{obs} against [OH$^-$] as illustrated in Fig. 4.12.

The kinetics of the reaction are then investigated at a constant alkaline pH using a buffer solution made up from a base and its conjugate acid. For example, this could be an amine (the base) and corresponding ammonium perchlorate (its conjugate acid). Buffer solutions of the same buffer *ratio* have the same p*H*, i.e. the same hydroxide ion concentration; by using different buffer *concentrations*, the effect of the concentration of the general base (the amine) can be investigated. If k_{obs} is independent of the buffer concentration as illustrated in Fig. 4.13 (where results at two p*H*s are included), then the reaction is confirmed as specific base catalysed.

In Fig. 4.13, the buffer ratio r1 (defined as [conjugate acid]/[base] as in Fig. 4.5 on p. 53) is lower than buffer ratio r2, i.e. the p*H* is higher in the solution of buffer ratio r1. In the context of base catalysis, it is quite in order to redefine the buffer ratio as

r = [base]/[conjugate acid]

in which case the relationship between p*H*, pK_a, and r becomes

p*H* = pK_a + log r .

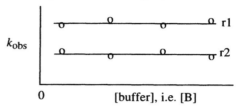

Fig. 4.13 Graphs of k_{obs} against [buffer] at two values of r for specific base catalysis

Specific base catalysis and reaction mechanisms. We saw earlier that pre-equilibrium protonation of substrate followed by a rate-limiting step which does not involve proton transfer leads to a specific acid catalysis rate law. Correspondingly, a reaction which involves pre-equilibrium *deprotonation* of substrate followed by a rate-limiting step which does not involve proton transfer necessarily leads to a specific base catalysis rate law. The Aldol reaction in dilute solution is a familiar organic reaction of this type, eqn 4.12.

The alkaline hydrolysis of simple esters (the B2 mechanism) is also first order in hydroxide but, since hydroxide also reacts with the carboxylic acid which is formed, it is not catalytic. The reaction is usually carried out with a large excess of hydroxide over ester so, in practice, the hydroxide concentration remains approximately constant.

$$2 \quad \underset{CH_3}{\overset{H}{\underset{\text{\Large C}}{|}}} \hspace{-0.3em} \overset{\text{O}}{\diagdown} \quad \xrightarrow[\text{H}_2\text{O}]{\text{OH}^-,\text{B}} \quad \underset{CH_3}{\overset{OH}{\underset{\text{CH}}{|}}} \hspace{-0.3em} \underset{}{\overset{H}{\underset{\text{C}}{|}}} \hspace{-0.3em} \overset{\text{O}}{\diagdown} \qquad (4.12)$$

Experimentally, the rate law for this reaction carried out in the presence of hydroxide and a general base, B, is

$$\text{rate} = k[\text{OH}^-][\text{CH}_3\text{CHO}]^2$$

i.e.   there is no uncatalysed reaction (no solvent-induced k_o reaction path), the reaction is first order in [OH⁻], it is independent of [B] at constant p*H*, it is second order in ethanal,

If the Aldol reaction is carried out at constant hydroxide concentration, the reaction becomes *pseudo* second order in ethanal, i.e.   rate = k_{obs} [CH₃CHO]² where $k_{obs} = k$ [OH⁻]

and   the overall reaction rate constant, *k*, is third order.

A mechanism for the reaction of eqn 4.12 involving pre-equilibrium deprotonation by any base present, and rate-limiting addition of the enolate carbanion so formed to the carbonyl of another ethanal molecule is shown in Fig. 4.14. Note in particular that both hydroxide and B are involved in deprotonation pre-equilibria, and that the equilibrium constants *K* and *K'* which characterise these equilibria will be maintained throughout the reaction. Correspondingly, two equilibria follow the rate-limiting step to generate the product; the proton donors are water and the conjugate acid of B (BH⁺) formed in one of the initial pre-equilibria. These two steps regenerate OH⁻ and B, so the proposed mechanism is fully catalytic.

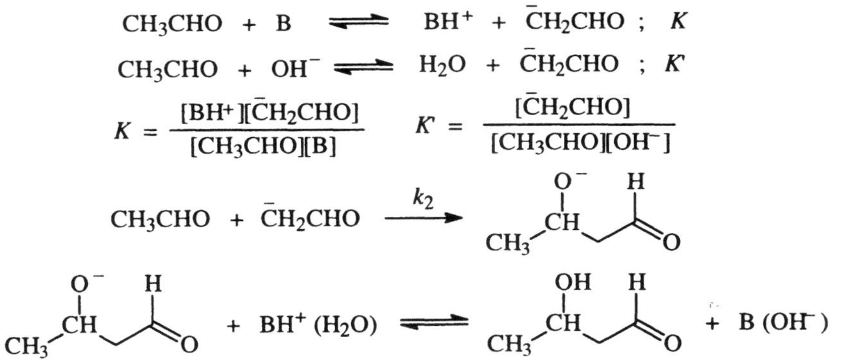

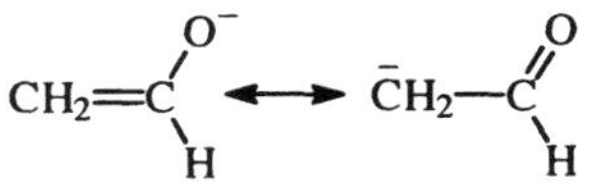

Fig. 4.14 Mechanistic proposal for the Aldol reaction in dilute aqueous basic solution

According to the mechanistic proposal in Fig. 4.14, the rate of the reaction is determined by the bimolecular k_2 step,

$$rate = k_2[CH_3CHO][CH_2CHO^-]$$

but we do not know the concentration of the enolate carbanion. However, it is related to the concentrations of starting materials by two equilibrium constants, K and K'. For convenience, we use the latter hence

$$rate = k_2[CH_3CHO] \, K' \, [CH_3CHO][OH^-] \, .$$

By combining like terms, this may be rewritten

$$rate = k_2 \, K' \, [CH_3CHO]^2 \, [OH^-] \, .$$

The mechanism of Fig. 4.14, therefore, leads to a rate law which has exactly the same form as that which is observed experimentally, and the experimental third-order rate constant, k, corresponds to the product of constants in the rate law derived from the mechanism:

$$k = k_2 \, K' \, .$$

Note that the general base B is involved in the mechanism, but it does not occur in the rate law.

4.7 General base catalysis

The Aldol reaction described above refers to experimental conditions of low ethanal concentration, and the rate-limiting step is the bimolecular reaction between an enolate carbanion with another ethanal molecule. If the initial concentration of ethanal is much higher, the bimolecular k_2 step will be faster and could compete with the reverse of the first step. If the second step becomes so fast that the carbanion is trapped by an aldehyde molecule as soon as it is formed, the reverse of the first step is prevented; the initial proton abstractions are now irreversible and rate limiting. This mechanistic possibility is illustrated in Fig. 4.15.

Water is not included in the expression for K' because it is the solvent (as well as a component in the reaction) so its activity is related to the mole fraction scale and hence is unity.

Any mechanism involving a pre-equilibrium deprotonation followed by a rate-determining step which does not involve proton transfer will lead to a specific base catalysis rate law.

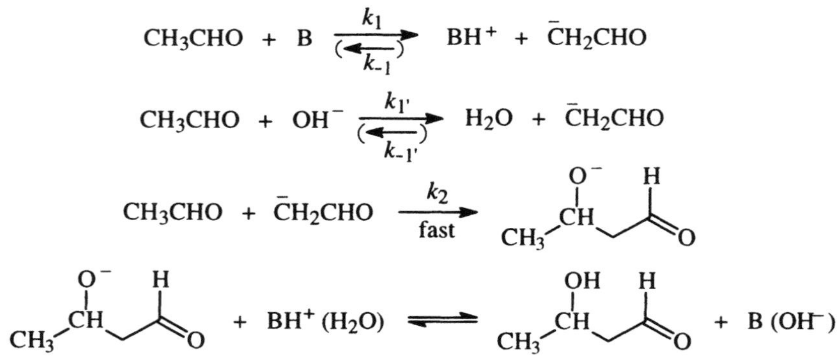

Fig. 4.15 Mechanistic proposal for the Aldol reaction at high ethanal concentration

We can now consider the kinetic consequences of this mechanistic changeover by working out what the new rate law will be. If formation of the reactive intermediate, i.e. deprotonation of the aldehyde, is rate-limiting, the rate of the reaction will be the sum of the rates of all routes by which the enolate is formed:

$$\text{rate} = k_1[CH_3CHO][B] + k_{1'}[CH_3CHO][OH^-]$$

$$\text{rate} = (k_1[B] + k_{1'}[OH^-])[CH_3CHO]$$

$$\text{rate} = k_{obs}[CH_3CHO] \text{ at constant } [B] \text{ and } [OH^-]$$

where $k_{obs} = k_1[B] + k_{1'}[OH^-]$.   (4.13)

Note that now, according to this mechanistic proposal,

there is still no solvent-induced (uncatalysed) reaction (no k_o path),

the reaction is first order in ethanal,

at constant [B], the reaction is first order in [OH⁻],

at constant pH, the reaction is first order in [B],

and k_{obs} is now a *pseudo* first-order rate constant.

For high concentrations of aqueous ethanal in the absence of a general base, a reaction *pseudo* first-order in ethanal is observed and the rate constant increases linearly with hydroxide concentration. In buffered alkaline solutions, the *pseudo* first-order rate constant is linearly dependent upon [B]. Thus, the experimental outcome is in accord with eqn 4.13 and the mechanism of Fig. 4.15. If further bases, B', B'', etc., are present, this mechanism predicts an additional term in the rate law of eqn 4.13 for each.

> Any mechanism in solution involving rate-determining proton abstraction will lead to a general base catalysis rate law.

Experimental characterisation of a general base catalysed reaction

The conversion of reactants R into products P in eqn 4.14 is catalysed by any base, B, as well as by OH⁻ and has an uncatalysed reaction channel.

$$R \xrightarrow[\text{H}_2\text{O}]{\text{B, OH}^-} P \qquad (4.14)$$

Assuming the reaction is first order in [R], we may write the rate law

$$-d[\text{R}]/dt = k_{\text{obs}}[\text{R}] \qquad \text{at constant [OH}^-] \text{ and [B]}$$

where $\qquad k_{\text{obs}} = k_0 + k_{\text{OH}}[\text{OH}^-] + k_{\text{B}}[\text{B}]$. $\qquad (4.15)$

When the reaction of eqn 4.14 is investigated in solutions buffered by combinations of the catalytic base B and its conjugate acid BH^+, the relationship between pH, the pK_a of BH^+, and the buffer ratio r is given by

$$\text{p}H = \text{p}K_a + \log r \qquad \text{where } r = [\text{B}]/[\text{BH}^+] .$$

Consequently, by maintaining the buffer ratio constant, we see it is straightforward to make up solutions of different [B] but identical pH.

An easy way to buffer an uncharged organic base such as an amine in water is simply to add a strong mineral acid, e.g. perchloric acid, until the desired pH is obtained. If the initial solution of the base is sufficiently concentrated, the buffered solution may then be diluted to obtain a set of solutions of the same buffer ratio (same pH).

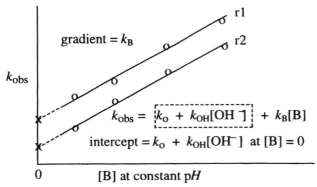

Fig. 4.16 Sketch of graphs of k_{obs} against general base concentration at buffer ratios r1 and r2 for a general base catalysed reaction

The procedure, therefore, is to measure the experimental *pseudo* first-order rate constant, k_{obs}, initially in solutions containing only hydroxide, i.e. [B] = 0 in eqn 4.15. By treating the reaction as though it is specific base catalysed, k_0 and k_{OH} are determined from the intercept and gradient of a plot of k_{obs} against [OH$^-$] as in Fig. 4.12. Next, values of k_{obs} are measured at constant pH in a set of solutions containing the general base buffered by its conjugate acid at constant buffer ratio but different buffer concentrations. This allows the construction of a graph of k_{obs} against [B] as in Fig. 4.16 from which the intercept at constant pH is the sum of the two terms in the dotted rectangle, and the gradient is k_{B}. At a different buffer ratio, a different intercept will be obtained corresponding to a different [OH$^-$] (but with the same k_0 and k_{OH}); the gradient, however, will be the same.

General base catalysis and reaction mechanisms

The experimental observation of a general base catalysis rate law is often mechanistically ambiguous because several mechanisms, besides rate-limiting proton abstraction (as in the Aldol reaction at high concentration), may lead to the same rate law. The hydrolysis of reactive esters such as ethyl dichloroethanoate and 4-nitrophenyl ethanoate, for example, are reactions which have general base catalysis rate laws, but cannot involve rate-limiting proton abstraction from the substrate. Typical bases in such reactions are amines like pyridine or imidazole (buffered with pyridinium or imidazolium perchlorate, for example) or sodium ethanoate (buffered with ethanoic acid).

Nucleophile catalysis. We have already encountered nucleophile catalysis, one of the two main hydrolytic mechanisms which lead to the general base catalysis rate law. We shall illustrate the mechanism in the present context with the hydrolysis of 4-nitrophenyl ethanoate catalysed by ethanoate.

This reaction was investigated in solutions buffered just on the acid side of neutrality in order to suppress the rapid hydrolysis induced by hydroxide (but it was necessary to establish that ethanoic acid itself is not catalytically active). This illustrates there is no paradox about having general base catalysis under acidic conditions.

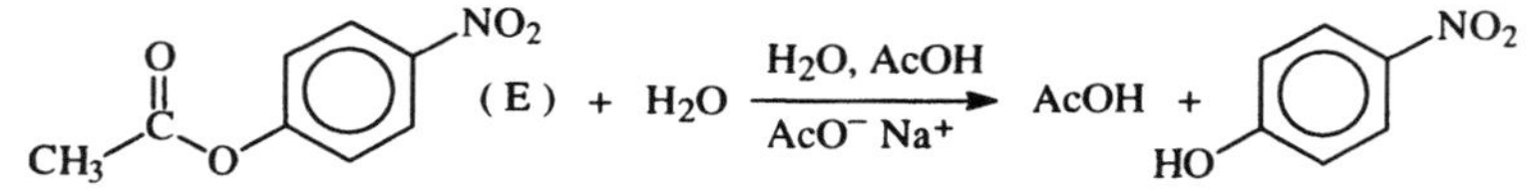

(4.16)

Since the solution is acidic, the intercept in Fig. 4.17 will include some reaction catalysed by the small concentration of H_3O^+, i.e., $k_o = k_o' + k_H[H_3O^+]$ where k_o' is the rate constant for the reaction genuinely induced by just water, and k_H the catalytic constant for reaction catalysed by hydronium ion.

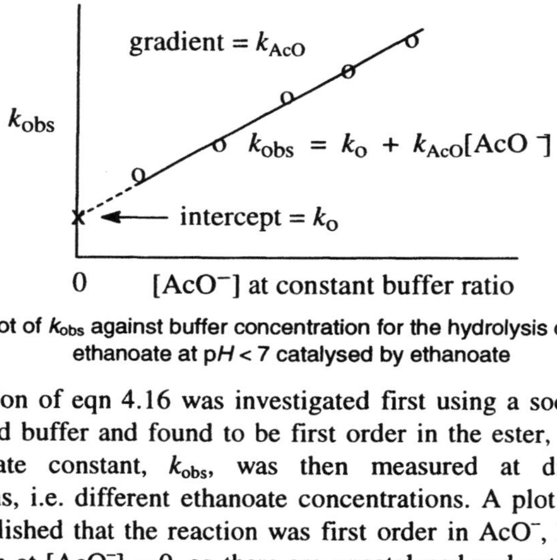

Fig. 4.17 Plot of k_{obs} against buffer concentration for the hydrolysis of 4-nitrophenyl ethanoate at p*H* < 7 catalysed by ethanoate

The reaction of eqn 4.16 was investigated first using a sodium ethanoate /ethanoic acid buffer and found to be first order in the ester, E. The *pseudo* first-order rate constant, k_{obs}, was then measured at different buffer concentrations, i.e. different ethanoate concentrations. A plot of k_{obs} against $[AcO^-]$ established that the reaction was first order in AcO^-, Fig. 4.17, with some reaction at $[AcO^-] = 0$, so there are uncatalysed and catalysed reaction channels and the rate law may be written

$$-d[E]/dt = k_{obs}[E] \qquad \text{where} \qquad k_{obs} = k_o + k_{AcO}[AcO^-] .$$

A reasonable mechanistic starting point is to assume that the uncatalysed reaction is the same as that for simple esters, and that the catalysed channel involves ethanoate in the way that hydroxide is involved in alkaline hydrolysis, i.e. as a nucleophile. Following formation of a tetrahedral intermediate, ethanoic anhydride is formed which, under the conditions of the reaction, will react further (via another tetrahedral intermediate) as shown in Fig. 4.18.

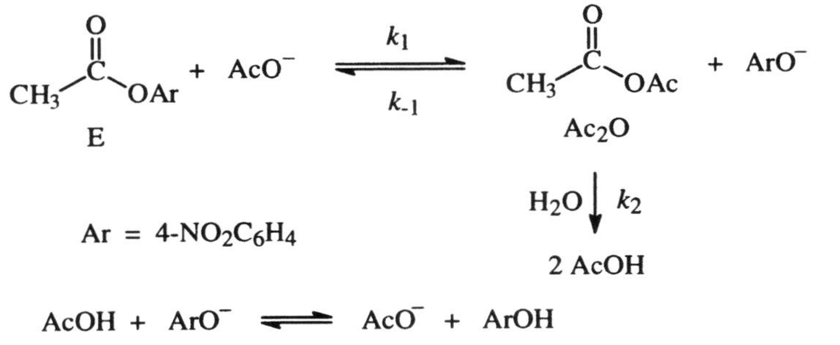

Ar = 4-NO$_2$C$_6$H$_4$

$$AcOH + ArO^- \rightleftharpoons AcO^- + ArOH$$

Fig. 4.18 A mechanism for the hydrolysis of 4-nitrophenyl ethanoate catalysed by ethanoate

Hydrolysis of one molecule of the ethanoic anhydride intermediate gives two molecules of ethanoic acid. However, ethanoic acid is a stronger acid than 4-nitrophenol, so a post-equilibrium involving the 4-nitrophenolate liberated in the first step effectively removes one ethanoic acid molecule to regenerate the ethanoate catalyst.

We now need to derive the rate law for the catalytic mechanism in Fig. 4.18 which involves ethanoic anhydride (Ac$_2$O) as a reactive intermediate, and see how this compares with what is observed experimentally. According to this mechanism, the rate of the ethanoate-catalysed reaction channel is given by eqn 4.17. However, this is not a legitimate rate law as it is not in terms of just reactant and catalyst.

$$d[AcOH]/dt = k_2[Ac_2O][H_2O] \qquad (4.17)$$

One way forward is to substitute for [Ac$_2$O] using the steady state approximation, i.e. assume that the rate of formation of the reactive intermediate in the catalysed reaction is just balanced by its rate of decomposition:

$$k_1[E][AcO^-] = k_{-1}[Ac_2O][ArO^-] + k_2[Ac_2O][H_2O]$$

$$k_1[E][AcO^-] = [Ac_2O](k_{-1}[ArO^-] + k_2[H_2O])$$

consequently, $$[Ac_2O] = \frac{k_1[E][AcO^-]}{k_{-1}[ArO^-] + k_2[H_2O]} .$$

Using this to substitute for [Ac$_2$O] in eqn 4.17 gives

$$\text{catalytic rate} = k_2.[H_2O]\frac{k_1[E][AcO^-]}{(k_{-1}[ArO^-] + k_2[H_2O])} \qquad (4.18)$$

which does *not* appear to be what is found experimentally.

If we assume that the reactive intermediate in the catalysed channel is captured by a solvent water molecule much faster than it reacts with a nitrophenoxide anion,

i.e. $$k_2[H_2O] \gg k_{-1}[ArO^-],$$

then eqn 4.18 may be simplified to give

$$\text{catalytic rate} = k_2.[H_2O]\frac{k_1[E][AcO^-]}{k_2[H_2O]}$$

or $\text{catalytic rate} = k_1[E][AcO^-]$.

The catalytic mechanism of Fig. 4.18 involves rate-limiting formation of the reactive intermediate, and the rate law derived from it has the same form as the catalytic term in the experimental rate law. If the mechanism is correct, the experimental catalytic rate constant is identical with the elementary rate constant of the first forward step in Fig. 4.18:

$$k_{AcO} = k_1 .$$

It was already known that aniline reacts with acetic anhydride in dilute aqueous solution to give acetanilde. In principle, however, the acetanilide could also have been formed through a direct reaction between aniline and the ester. In that event, the aniline would have led to a rate enhancement, i.e. there would have been a term in the rate law first order in [aniline]. No such rate effect was detected.

Demonstrating that the proposed mechanism leads to the observed rate law, however, does not prove that the mechanism is correct. Corroborative evidence is required–preferably direct evidence of ethanoic anhydride as a reactive intermediate. When the reaction was carried out in the presence of a low concentration of aniline ($PhNH_2$), acetanilide (PhNHAc) was formed. The most economical explanation of this product analysis evidence is that aniline, a much better nucleophile than water, traps some of the low concentration of ethanoic anhydride formed as an intermediate in the mechanism of Fig. 4.18.

Mechanistic general base catalysis. Ethanoic anhydride, the compound which was formed as a reactive intermediate in the ethanoate-catalysed hydrolysis of 4-nitrophenyl ethanoate discussed above, is a stable compound under non-hydrolytic conditions. Its own hydrolysis has been investigated and also shown to be catalysed by ethanoate, eqn 4.19.

Again, the reaction was buffered to be on the (slightly) acidic side of neutrality to ensure that there were no complications due to the much faster hydroxide-induced reaction.

$$(4.19)$$

The reaction is first order in ethanoic anhydride. The experimental rate law was investigated in the same way as that for 4-nitrophenyl ethanoate and is shown below. There is both a solvent induced reaction and a route catalysed by ethanoate:

$$\text{rate} = k_{obs}[Ac_2O]$$

where $k_{obs} = k_o + k_{AcO}[AcO^-].$

However, since nucleophilic attack by ethanoate at ethanoic anhydride can only lead to the formation of another molecule of ethanoic anhydride, this cannot possibly provide a catalytic mechanism. Exactly how ethanoate exerted its catalytic effect was something of a conundrum. Ethanoate is now thought to catalyse the reaction by enhancing the nucleophilicity of water. In other words, water alone hydrolyses ethanoic anhydride by acting as a nucleophile, but ethanoate ions (like some other general bases) also provide an additional route by enhancing the nucleophilicity of other water molecules through hydrogen bonding. The catalytic mechanism is shown in Fig. 4.19.

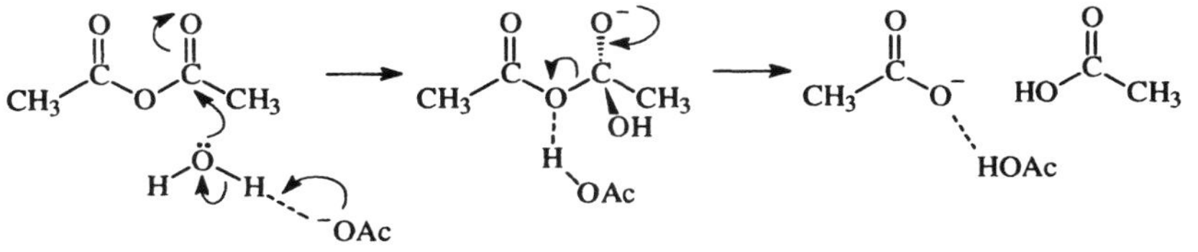

In this mechanism, a water molecule hydrogen bonded to ethanoate is the nucleophile, and this is more nucleophilic than a water molecule in bulk water. The ethanoic acid molecule generated in the formation of the tetrahedral intermediate can then hydrogen bond to the nascent nucleofuge and facilitate its departure.

Fig. 4.19 Proposed mechanism for hydrolysis of ethanoic anhydride catalysed by ethanoate

According to this mechanism, the rate of the catalysed channel will be given by:

$$\text{catalysed reaction rate} = k[H_2O][AcO^-][Ac_2O]$$

or, by combining constant terms,

$$= k'[AcO^-][Ac_2O]$$

which has the same form as the experimentally observed catalytic term.

The mechanism in Fig. 4.19 whereby the base exerts its catalytic effect by enhancing the nucleophilicity of the solvent, generally water, is commonly referred to as *general base catalysis*. This phrase is also used to describe an experimental finding, i.e. the rate law of eqn 4.15 on p. 63 or, in its generalised form,

$$k_{obs} = k_o + k_{OH}[OH^-] + \Sigma k_B[B]. \tag{4.20}$$

To reduce the confusion, we shall refer to the mechanism of Fig. 4.19 as *mechanistic general base catalysis*, and use the term *kinetic general base catalysis* for reactions which have the rate law of eqn 4.20 regardless of mechanism.

An intramolecular example of this mechanism involving an alcoholic OH as the assisted nucleophile is shown on p. 93, i.e. when the assisted nucleophile is not a solvent molecule.

Distinguishing between nucleophile catalysis and mechanistic general base catalysis for solvolytic reactions

We have already seen that kinetic general base catalysis is mechanistically ambiguous. The most obvious mechanism leading to this rate law is when deprotonation of the substrate is rate limiting, e.g. the Aldol reaction in concentrated solution. Generally, rate-limiting proton transfer is easy to characterise by substrate deuterium kinetic isotope effect measurements. If proton transfer is rate limiting, the deuterium analogue reacts appreciably slower than the protium one as we shall see in the next chapter. However, when the reaction does not involve rate-limiting deprotonation of the substrate but is solvolytic, the two commonest mechanisms leading to kinetic general base catalysis are nucleophile catalysis and mechanistic general base catalysis. It is not always easy to distinguish between them, but there are three general strategies, the first two being more widely applicable.

A solvolytic reaction is one in which the solvent is both the reaction medium and a reactant.
For example,
in water, it is hydrolysis,
in methanol, it is methanolysis,
in ethanoic acid (acetic acid) it is called acetolysis.
Solvolytic reactions are invariably *pseudo* first order in the substrate.

1. *Identification of an intermediate to confirm nucleophile catalysis.* We saw above in the hydrolysis of 4-nitrophenyl ethanoate that the intermediate, ethanoic anhydride, could be intercepted by aniline. This is an example of a general technique and, if the outcome is positive, i.e. an intermediate is trapped, provides incontrovertible evidence that the mechanism is nucleophile catalysis.

Alternatively, an intermediate may be confirmed spectroscopically as in the hydrolysis of 4-nitrophenyl ethanoate catalysed by imidazole (buffered with imidazolium perchlorate), eqn 4.21. In this reaction, the ester suffers nucleophilic attack by imidazole to give N-acetylimidazolium as a reactive intermediate in rapid equilibrium with its deprotonated form, N-acetylimidazole, which can be detected by UV spectroscopy.

N-acetylimidazole

$$\underset{CH_3}{} \quad + \; H_2O \; \xrightarrow[\substack{Im \\ ImH^+ \; ClO_4^-}]{H_2O} \; AcOH \; + \quad \qquad (4.21)$$

Failure to intercept a reactive electrophilic intermediate with a nucleophilic reagent, or to detect it spectroscopically, does not prove there is no intermediate. The mechanism may still involve nucleophile catalysis but via an intermediate too short-lived to be trapped or present in the reaction at too low a steady-state concentration to be detected spectroscopically.

2. *Solvent deuterium kinetic isotope effect measurements.* If the reaction of eqn 4.21 above is carried out in D_2O instead of H_2O, the rate is unchanged,

i.e. $\qquad\qquad k_{Im}(H_2O)/k_{Im}(D_2O) = 1.0$.

Reactions which, on the basis of other evidence, proceed via nucleophile catalysis, i.e. do not involve cleavage of a solvent O–L bond in the rate-limiting step, generally show H_2O/D_2O rate ratios of between 1 and 2.

In contrast, hydrolyses reacting via mechanistic general base catalysis give higher H_2O/D_2O rate ratios, usually between about 2 and 4. An example is the imidazole-catalysed hydrolysis of ethyl dichloroethanoate, eqn 4.22 (where L = H or D) for which $k_{Im}(H_2O)/k_{Im}(D_2O) = 3.0$.

L is used to represent an unspecified isotope of hydrogen.

$$Cl_2CHCO_2Et \; + \; L_2O \; \xrightarrow[Im/ImL^+ \; ClO_4^-]{L_2O} \; Cl_2CHCO_2L \; + \; EtOL \quad (4.22)$$

As shown earlier in Fig. 4.19, mechanistic general base catalysis involves the base accepting L^+ from a water molecule as the water acts as a nucleophile with the substrate in the rate-limiting step of the reaction. Consequently, to anticipate what will be considered in more detail in the next chapter, the reaction is appreciably slower when the transfer is of $^2H^+$ than of $^1H^+$.

When the proton transfer to the base is concerted with the nucleophilic attack of the water molecule, the rate law is kinetic general base catalysis. If a general base deprotonates water in a pre-equilibrium, then hydroxide acts as the nucleophile, a specific base catalysis rate law is observed.

3. *Measurement of the Brønsted coefficient, β.* The Brønsted coefficient is the parameter of a kinetic general base catalysed reaction which expresses the relationship between the catalytic effectiveness of a range of catalysts and

their strengths as bases, as described in the next section. It is the gradient of the plot of $\log k_B$ against $pK_a(BH^+)$, and ranges between 0 and 1 for reactions which involve rate-limiting proton transfer. However, it may be larger than unity for nucleophile catalysed reactions; consequently, such a value is often good evidence for nucleophile catalysis. This third strategy is only available if it is possible to measure k_B values for a range of catalysts, i.e. it cannot help to identify as nucleophile catalysed or mechanistic general base catalysed the mechanism of a hydrolytic reaction with a single catalyst.

4.8 The Brønsted relationship for catalysed reactions

We saw earlier that the origin of the hydronium ion in a specific acid catalysed reaction is irrelevant:

$$k_{obs} = k_o + k_H[H_3O^+] .$$

The catalytic constant k_H (like k_o) is a parameter of the catalysed reaction and has a particular value for that reaction under specified reaction conditions.

Correspondingly, k_{OH} in a specific base catalysed reaction with the rate law

$$k_{obs} = k_o + k_{OH}[OH^-]$$

has a particular value under particular conditions, and it is of no consequence whether the hydroxide comes from dissolved sodium hydroxide or is generated in a pre-equilibrium from the aqueous solvent by, for example, an amine, B:

$$B + H_2O = BH^+ + OH^- .$$

However, if a reaction is general acid catalysed:

$$k_{obs} = k_o + k_H[H_3O^+] + \Sigma k_{AH}[AH]$$

or general base catalysed:

$$k_{obs} = k_o + k_{OH}[OH^-] + \Sigma k_B[B] ,$$

then each catalyst has its own catalytic constant for that reaction. A particular general acid or base will probably catalyse a wide range of reactions, but the values of its catalytic constants for different reactions will be different and unrelated. Frequently, however, the catalytic constants of a *family* of general acids which catalyse the *same* general acid catalysed reaction are related; correspondingly, the catalytic constants of a *family* of bases which catalyse the *same* general base catalysed reaction are frequently related. These relationships are expressed in the celebrated Brønsted equation which is of central importance in the study of catalysis.

The Brønsted relationship for general acid catalysed reactions

It seems intuitively reasonable that, for a general acid catalysed reaction represented by eqn 4.4 below, stronger undissociated acids will be better catalysts than weaker ones, i.e. if AH is a stronger acid than A'H [$pK_a(AH) <$ $pK_a(A'H)$], then $k_{AH} > k_{A'H}$.

$$\text{Substrate} \xrightarrow[k_{obs}]{AH,\ H_3O^+,\ H_2O} \text{Products} \qquad (4.4)$$

$$k_{obs} = k_o + k_H[H_3O^+] + k_{AH}[AH],$$

Any structural modification to the catalytic acid AH which makes it a better acid, i.e. a better proton donor, should make it a more effective catalyst since acid catalysis involves proton transfer.

This is commonly observed and, for a particular reaction, there is frequently a quantitative relationship between the K_a values of a family of acid catalysts, AH, and their catalytic constants, k_{AH}. The relationship is logarithmic:

$$\log k_{AH} \propto \log K_a(AH)$$

or

$$\log k_{AH} = \alpha \log K_a(AH) + \text{constant}$$

where α is the constant of proportionality.

The differential version of this equation eliminates the second constant:

$$\delta(\log k_{AH}) = \alpha\ \delta(\log K_a)$$

$$\text{or } \delta(\log k_{AH}) = -\alpha\ \delta(pK_a) \qquad (4.23)$$

where, for convenience, we write just K_a rather than $K_a(AH)$.

Equation 4.23 is a version of the Brønsted equation (or relationship) for general acid catalysed reactions, and α is called the Brønsted coefficient (or Brønsted parameter). The value of α is a characteristic of the particular general acid catalysed reaction under specified conditions.

Note that acid strength is a matter of thermodynamics whereas catalysis is essentially a matter of reaction kinetics. The Brønsted equation, therefore, like the Hammett equation for rates considered earlier, is a rate–equilibrium correlation.
Note also that the Brønsted equation, like the Hammett equation, involves logarithms to the base 10.

Determination of α for a general acid catalysed reaction. The catalytic constant, k_{AH}, for a catalytic acid AH of known pK_a is determined for the chemical reaction by the methods already described. This is repeated for further catalytic acids A'H, A''H, etc. The $\log k_{AH}$ values are then plotted against the corresponding pK_a values, and the α value for this particular general acid catalysed reaction is the negative of the gradient as sketched in Fig. 4.20 for three different reactions.

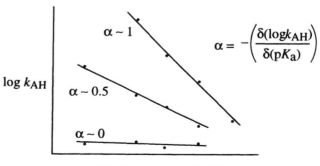

$$\alpha = -\left(\frac{\delta(\log k_{AH})}{\delta(pK_a)}\right)$$

Fig. 4.20 Brønsted plots for three different general acid catalysed reactions

Empirically, the plot with gradient ~0 corresponds to a general acid catalysed reaction for which catalytic acids of widely different pK_a values have very similar catalytic constants. The one with gradient ~-1 (α ~+1) corresponds to a general acid catalysed reaction whose rate is very sensitive to the acid strength of the catalytic acid, i.e. a 10 fold increase in acid strength of the catalyst [$\delta(pK_a) = -1$] leads to an approximately 10 fold increase in the catalytic constant [$\delta(\log k_{AH}) = +1$]. Between these two extremes in Fig. 4.20 is a Brønsted plot in which only about half of an increase in acid strength of the catalyst is expressed in its increased catalytic effectiveness.

The dehydration of ethanal hydrate (ethane-1,1-diol) in aqueous acetone is a general acid catalysed reaction with $\alpha = 0.54$:

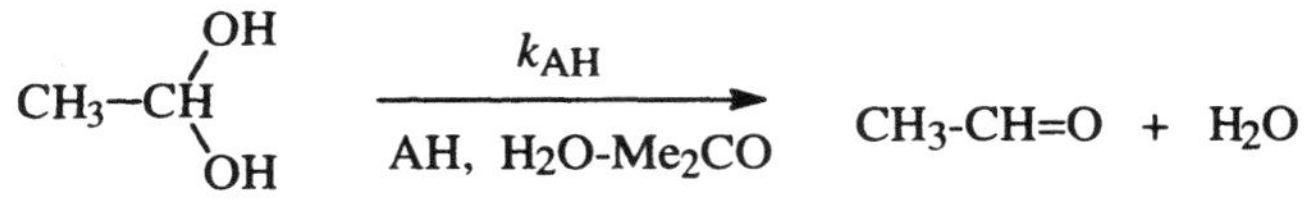

$$CH_3-\overset{\displaystyle OH}{\underset{\displaystyle OH}{CH}} \quad \xrightarrow[AH, \ H_2O-Me_2CO]{k_{AH}} \quad CH_3\text{-}CH\text{=}O \ + \ H_2O$$

According to simple theory, Brønsted α values should range between 0 and 1 but, in practice, it is quite difficult to measure values close to zero and close to unity; consequently, most published values are in the range 0.2 to 0.8.

Interpretation of Brønsted α values. Within the context of general acid catalysed chemical reactions which have rate-limiting proton transfer, the magnitude of the Brønsted coefficient is a measure of the extent of proton transfer from catalyst, A-H, to substrate, S, in the transition state on the way to product, P, eqn 4.24 and Fig. 4.21.

$$S \ + \ H\text{-}A \ \rightarrow \ (SHA)^{\ddagger} \ \rightarrow \ P \ + \ H\text{-}A \qquad (4.24)$$

If the proton is hardly transferred in the transition state, then a poor proton donor will be as good a catalyst as a much better one, and α will be small. At the other extreme, if the proton is almost completely transferred from catalyst to substrate in the transition state, then a stronger acid, i.e. better proton donor, will be a much more effective catalyst, and α will be close to unity. A reaction in which the proton is about half transferred will lead to an α value of about 0.5. In practice, the α value is measured experimentally and its value provides information about the nature of the transition state.

An α value close to zero corresponds to an acid catalysed reaction for which weak acids are all comparably effective. The reaction, therefore, will be dominated by the proton donor present in highest concentration, i.e. water when the reaction is in aqueous solution. Catalysis is difficult to detect for such a reaction–it will appear uncatalysed.

An α value close to unity corresponds to an acid catalysed reaction dominated by the strongest acid present. In water, the strongest acid possible is hydronium ion, so such reactions appear to be specific acid catalysed.

$\left(S\text{-----}H^{\cdots}A\right)^{\ddagger}$	$\left(S\text{----}H\text{----}A\right)^{\ddagger}$	$\left(S^{\cdots}H\text{-----}A\right)^{\ddagger}$
low extent of proton transfer in transition structure	proton about half transferred in transition structure	proton transfer almost complete in transition structure
$\alpha \rightarrow 0$	$\alpha \sim 0.5$	$\alpha \rightarrow 1$

Fig. 4.21 Representations of transition structures for the general acid catalysed reaction of eqn 4.24 with Brønsted α-values approximately 0, 0.5, and 1

The Brønsted relationship for general base catalysed reactions

Parallel with ideas expressed above for general acid catalysed reactions, we may reasonably expect that the stronger a compound is as a base, the better it will be as a catalyst for a general base catalysed reaction, eqn 4.25. In fact, it was on base catalysed reactions that Brønsted first carried out his pioneering experimental studies on catalysis, and it was on such reactions that he established the equation which today bears his name.

$$\text{Reactant} \xrightarrow[k_B]{\text{B, H}_2\text{O}} \text{Product} \qquad (4.25)$$

For a family of bases B, B', B", etc., which catalyse a particular reaction, it is commonly observed that there is a quantitative relationship between the strength of the base and its catalytic constant, k_B, and the relationship is logarithmic. This corresponds to an inverse relationship between the catalytic constant of the base and the dissociation constant of the conjugate acid of the base.

In logarithmic form,

$$\log k_B \propto -\log K_a(\text{BH}^+)$$

According to modern convention, the strength of a base, B, is expressed in terms of the acid strength of its conjugate acid, BH^+:

$BH^+ + H_2O = B + H_3O^+$; $K_a(BH^+)$

$$K_a(\text{BH}^+) = \frac{[\text{B}][\text{H}_3\text{O}^+]}{[\text{BH}^+]}$$

and $pK_a = -\log K_a$. Consequently, the stronger B is as a base, the larger the value of the pK_a of its conjugate acid, BH^+.

or $$\log k_B = -\beta \log K_a(\text{BH}^+) + \text{constant}$$

where β is the constant of proportionality,

or $$\log k_B = \beta\, pK_a(\text{BH}^+) + \text{constant}$$

so, in its differential form,

$$\delta(\log k_B) = \beta\, \delta(pK_a) \qquad (4.26)$$

where, for convenience, we write just pK_a rather than $pK_a(\text{BH}^+)$.

So, in spite of our convoluted means of arrival, the modern version of the Brønsted equation in eqn 4.26 for general base catalysed reactions is simple; the Brønsted coefficient, β, is a characteristic of the particular general base catalysed reaction in eqn 4.25 under specified conditions.

According to simple theory, β values may range between 0 and 1 for reactions with rate-limiting deprotonation. A β value approaching unity corresponds to a reaction for which the catalytic constants are maximally sensitive to the base strength of catalysts; a β value close to zero corresponds to a reaction for which the catalytic constants hardly increase at all as increasingly stronger bases are used as catalysts.

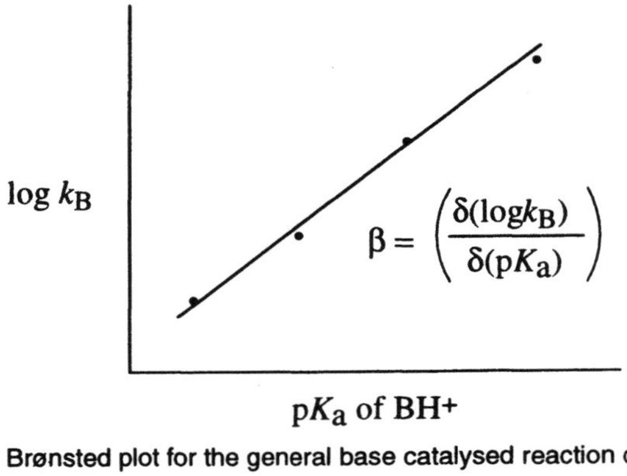

$$\beta = \left(\frac{\delta(\log k_B)}{\delta(pK_a)} \right)$$

Fig. 4.22 Brønsted plot for the general base catalysed reaction of eqn 4.25

Determination of β for a general base catalysed reaction. By the methods already described, the catalytic constants, k_B, for each of a range of general bases, e.g. a family of amines, for the reaction are measured. Logk_B values are then plotted against the pK_a values of the conjugate acids of the catalytic bases (the ammonium cations) as sketched in Fig. 4.22, and β is the gradient.

Interpretation of Brønsted β values. The magnitude of the Brønsted coefficient is taken as a measure of the extent of proton transfer in the transition state of a general base catalysed reaction which involves rate-limiting deprotonation of the substrate. This is exactly parallel with the interpretation of the Brønsted α in general acid catalysed reactions. A β value close to unity corresponds to a transition structure in which the proton transfer from substrate to catalyst is almost complete, i.e. a stronger base will be a better catalyst. A β value close to zero corresponds to a transition structure in which the proton transfer from substrate to catalyst has hardly begun, i.e. it makes little difference whether the catalyst is a strong base or not.

The Brønsted relationship for nucleophile catalysed and mechanistic general base catalysed hydrolytic reactions

We have already seen that the kinetic general base catalysis rate law for the hydrolysis of, for example, a reactive ester may be a consequence of either nucleophile catalysis or mechanistic general base catalysis. In the latter, the base catalyst accepts a proton from a water molecule as the water acts as nucleophile. Since proton transfer is involved in this reaction (though not from the substrate), the gradient of a plot of $\log k_B$ against $pK_a(BH^+)$ is constrained (at least by simple theory) to range between zero and unity, and the plot may be regarded as a conventional Brønsted correlation.

In the nucleophile catalysis mechanism, however, proton transfer is not involved at all; a plot of $\log k_B$ against $pK_a(BH^+)$ now simply expresses an empirical relationship between catalytic effectiveness and base strength for the family of nucleophile catalysts. In such a correlation, there is no reason in principle why the gradient, i.e. the Brønsted coefficient, should not be greater than unity. Indeed, observation of a Brønsted coefficient greater than unity for a hydrolysis reaction catalysed by a family of compounds which may act as nucleophiles or bases is evidence that the mechanism is nucleophile catalysis rather than mechanistic general base catalysis.

4.9 Brønsted-type correlations for uncatalysed nucleophilic substitution reactions—β_{nuc} and β_{lg}

So far in this chapter, we have focused exclusively on catalysed reactions. We have seen that the Brønsted relationship for a general acid or general base catalysed reaction is a logarithmic correlation between the catalytic rate constants of a family of catalysts and the acid or base strengths of the catalysts. Other logarithmic correlations between rate constants and equilibrium constants of exactly the same form as the Brønsted relationship are conveniently considered here even though they do not involve catalysis.

The Hammett equation for kinetics considered in chapter 3 is also a logarithmic correlation of rate constants and equilibrium constants.

Correlation of nucleophilicity with base strength

We saw in chapter 1 how group transfer reactions can be described and analysed using reaction maps. Here, we are considering how the mechanisms are actually investigated, i.e. how the experimental basis for the analysis is established.

Substitution at unsaturated carbon. Equation 4.27 represents a family of substitution reactions in which phenolate anions, $ZC_6H_4O^-$, displace an unspecified nucleofuge, X^-, from an acyl group. Each rate constant k_{nuc} for this reaction reflects the nucleophilicity of each phenolate with a common electrophile. Not surprisingly, the order of nucleophilicities as Z is changed along the family of phenolates (phenoxides) is the same as the order of their base strengths.

$$X^- + R\overset{+}{C}O + Y^- \qquad X^- + \overset{O}{\underset{R\,\,\,\,\,Y}{\|}}$$

$$\uparrow$$

C - X
bond
cleavage

$$\overset{O}{\underset{R\,\,\,\,\,X}{\|}} + Y^- \xrightarrow[\text{C - Y bond formation}]{} \quad R\overset{X}{\underset{Y}{\diagdown}} O^-$$

Reaction map template for acyl transfer

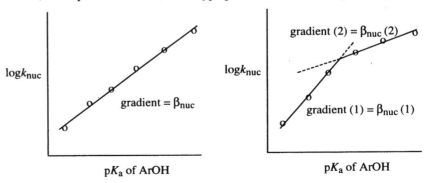

In fact, there is often a quantitative relationship between the two which is logarithmic and expressed in its differential form in eqn 4.28:

$$\delta(\log k_{nuc}) = \beta_{nuc}\, \delta(pK_a) \qquad (4.28)$$

where k_{nuc} = the rate constant for a reaction described by eqn 4.27,
  K_a = the acidity constant of the conjugate acid, ArOH, of the nucleophile, ArO⁻,
and β_{nuc} = the coefficient of this Brønsted-type correlation,
  i.e., the gradient of the plot of $\log k_{nuc}$ against pK_a.

β_{nuc} is a parameter characteristic of the family of reactions.

An example of the use of the β_{nuc} parameter is in helping to determine whether the family of reactions described by eqn 4.27 involve concerted displacements of X^- by ArO^- via single transition structures (4.1), or whether the reactions are step-wise, i.e. via tetrahedral intermediates (4.2). If the mechanisms are concerted, then substituents Z which affect the affinity of the phenolates for a proton, i.e. their base strengths, will have parallel effects upon the affinity of the phenolates for the electrophilic reactant RCOX. In this event, we expect a linear Brønsted-type plot as sketched in Fig. 4.23(a).

$$\left(\overset{O}{\underset{ArO^{\delta-}}{\overset{\|}{R-C\cdots\cdots X^{\delta-}}}} \right)^{\ddagger}$$

(4.1)

$$\left[\overset{X}{\underset{ArO\,\,\,\,R}{\overset{\diagup O^-}{C}}} \right]$$

(4.2)

Fig. 4.23 Brønsted-type plots for the reactions of eqn 4.27
(a) when the mechanism is concerted via transition structures (4.1)
(b) when the mechanism is step-wise via tetrahedral intermediates (4.2)

Acyl transfer between weakly basic anions shown in Fig. 4.24 where Z represents electron-withdrawing groups are concerted reactions of this type.

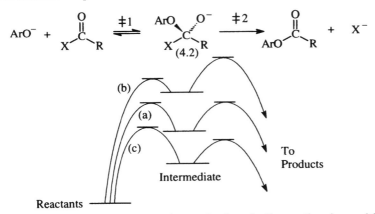

The nature of the R group in Fig. 4.24 affects the synchroneity of the concerted mechanism.

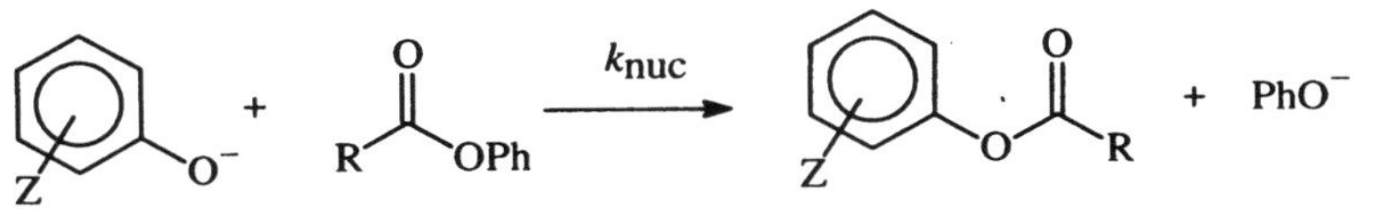

Fig. 4.24 Acyl transfer between weakly basic oxyanions with concerted mechanisms

If the mechanism of the reaction in eqn 4.27 is step-wise via tetrahedral intermediates (4.2) as shown in Fig. 4.25, the rate-limiting step could be either the formation of (4.2), or its decomposition. In profile (a) of Fig. 4.25, the phenoxide adds to give a tetrahedral intermediate which may continue to products or return to starting materials, and the free energy barriers indicate that these alternatives have comparable elementary rate constants. If the phenoxide is made a poorer nucleophile, e.g. by introducing a nitro group, then the first step will have a higher barrier and give a less stable intermediate. Profile (b) describes this mechanism and has been superimposed upon the original in Fig. 4.25 so that the initial states are at a common level in the free energy axis. The intermediate in (b) returns to starting materials faster than it continues to products.

Acyl transfers are normally step-wise i.e. concerted mechanisms are uncommon for this type of reaction.

These changes to the first step of the reaction may be considered in terms of parallel effects upon the structure and energy of the activated complex ‡1 (see chapter 1). To a first approximation, the barriers for the second steps in these three profiles are the same since they correspond to the departure of the same nucleofuge X^-. However, the nucleofuge departs from a different residue in the three cases which causes the barriers to be slightly different and in the order: (b) > (a) > (c).

Fig. 4.25 Reaction profiles for step-wise mechanisms for the reactions in eqn 4.27

Profile (c) for this step-wise mechanism, also superimposed with reactants at the same free energy level, is for the reaction of a phenoxide more nucleophilic than the original one in (a). This gives a more stable tetrahedral intermediate which proceeds to products faster than it returns to reactants.

For strongly nucleophilic phenoxides whose mechanisms are described by profiles like (c), the rate-limiting step is the initial addition, i.e., in the intermediate, X^- is a better nucleofuge than ArO^-. In such reactions, the β_{nuc} reflects the change between the initial state (in which the phenoxide bears a full negative charge) and transition state ‡1 (in which a partial bond has formed from the phenoxide to the electrophile). However, for weakly nucleophilic phenoxides whose mechanism is described by profiles like (b), i.e. when the original nucleophile is a better nucleofuge in the intermediate than X^-, the second step is rate limiting. The oxygen of the original phenoxide nucleophile is covalently bonded in the transition state of this step of the reaction, $\ddagger 2$. Consequently, the change between the very initial state and overall transition state for this oxygen will be appreciably greater than when the mechanism follows path (c), and a larger β_{nuc} is anticipated. We see, therefore, that in a step-wise mechanism for acyl transfer, we expect a Brønsted-type plot to have a larger β_{nuc} for less reactive nucleophiles and a smaller β_{nuc} for more reactive ones as in Fig. 4.23(b). The discontinuity occurs when the barrier heights of forward and back reactions from the tetrahedral intermediate are the same as in profile (a) of Fig. 4.25, e.g. when nucleofuge and nucleophile are identical ArO^- anions in eqn 4.27.

Bimolecular substitution at saturated carbon. In chapter 3, σ-values of substituents in phenoxide and carboxylate nucleophiles were used in Hammett correlations of S_N2 reactions.

$$Nuc^- \ + \ R\text{-}X \ \xrightarrow{\ k_{nuc}\ } \ R\text{-}Nuc \ + \ X^-$$

The sensitivity of reactions of a common electrophile to substituents in a family of aromatic nucleophiles was expressed as the ρ-value. Another method of correlation when pK_a data of the conjugate acids of the nucleophiles, e.g. amines, phenoxides, and carboxylates, are available is to use eqn 4.28. Whereas use of the Hammett equation is restricted to aromatic compounds, this Brønsted-type approach is not.

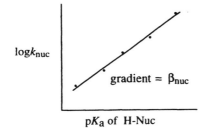

$\log k_{nuc}$

gradient = β_{nuc}

pK_a of H-Nuc

Correlation of nucleofugacity with base strength

Substitution at unsaturated carbon. In the reactions of eqn 4.29, a common nucleophile, X^-, displaces a series of phenoxides from the acyl residue. These are the reverse of the reactions in eqn 4.27 and so may also be step-wise or concerted.

$$X^- \ + \ \underset{R}{\overset{O}{\|}}{C}\text{-}O\text{-}\bigcirc_Z \ \xrightarrow{\ k_{lg}\ } \ \underset{R}{\overset{O}{\|}}{C}\text{-}X \ + \ ^-O\text{-}\bigcirc_Z \qquad (4.29)$$

We expect that more basic phenoxides will be poorer nucleofuges. This is generally observed, and eqn 4.30 expresses the relationship:

$$\delta(\log k_{lg}) \ = \ - \beta_{lg}\ \delta(pK_a) \qquad (4.30)$$

where k_{lg} = the rate constant for the reaction of eqn 4.29,
   K_a = the acidity constant of the conjugate acid, ArOH, of the nucleofuge, ArO^-, and
   β_{lg} = the coefficient of this Brønsted-type correlation,
      i.e. the negative of the gradient in the plot of $\log k_{lg}$ against pK_a.

β_{lg} is a parameter characteristic of the family of reactions.

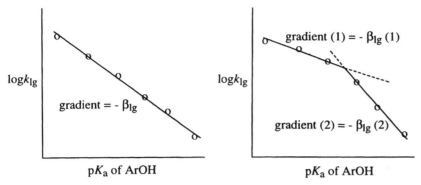

Fig. 4.26 Brønsted-type plots for the reactions of eqn 4.29
(a) when the mechanism is concerted via transition structures (4.1)
(b) when the mechanism is step-wise via tetrahedral intermediates (4.2)

If the reactions of eqn 4.27 have concerted mechanisms via transition structures (4.1), then their reverse, eqn 4.29, must also be concerted via the same transition structures, and will again lead to a simple linear Brønsted-type correlation. For the reactions of eqn 4.29, however, the gradient will be negative (leading to a positive β_{lg}) as sketched in Fig. 4.26(a).

Profiles for the step-wise possibilities for eqn 4.29 are shown in Fig. 4.27. The initial nucleophilic attack by X^- gives a tetrahedral intermediate which then leads on to product by expulsion of the nucleofuge ArO^-. The nucleofuge (which bears the substituent) remains covalently bonded in the intermediate, so the effects of substituents upon just this first step will be small as indicated in Fig. 4.27. The effects of substituents will be greater in the second step when the bond to the nucleofuge is being broken.

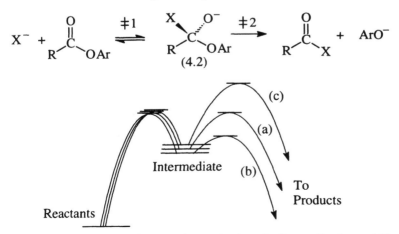

Fig. 4.27 Reaction profiles for step-wise mechanisms for the reaction in eqn 4.29

Profile (a) in Fig. 4.27 describes the reaction when X⁻ and ArO⁻ are comparably good nucleofuges, so the tetrahedral intermediate partitions equally between return to starting materials and forward reaction to products. Profile (b) describes the reaction when ArO⁻ is made a better nucleofuge than X⁻, e.g. by introduction of a nitro group into Ar, and the pK_a of its conjugate acid is smaller. The forward reaction from the intermediate is now faster than the reverse of the first step, so the first step is rate limiting in the overall reaction. Thirdly, profile (c) describes the reaction when ArO⁻ has been made a worse nucleofuge than X⁻; return to starting materials is faster than the second forward step which is now rate limiting in the overall reaction.

Consequently, the complete β_{lg} Brønsted-type correlation for step-wise mechanisms has a discontinuity as sketched in Fig. 4.26(b). There is a small β_{lg} for those faster reactions described by profile (b) in Fig. 4.27, i.e. with rate-limiting initial addition, and a larger value for the slower reactions when the second step, the departure of the nucleofuge, is rate limiting.

Substitution at saturated carbon. Correlations of nucleofugacity are possible using eqn 4.30 as illustrated below for a generic S_N2 reaction.

$$X^- \; + \; R\text{-Lg} \; \xrightarrow{\;k_{lg}\;} \; R\text{-X} \; + \; Lg^-$$

This method is not restricted to aromatic compounds in the way that use of the Hammett equation is, but pK_a values of the conjugate acids of the nucleofuges, H-Lg, must be known. In principle, it could also be used for S_N1 reactions but, in practice, substrates with nucleofuges with any basic tendencies seldom undergo S_N1 reactions; S_N1 reactions normally require nucleofuges which, under the conditions of the reaction, are virtually non-basic.

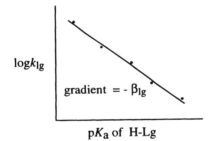

$\log k_{lg}$

gradient $= - \beta_{lg}$

pK_a of H-Lg

Problems

4.1 The Benzoin condensation is catalysed by cyanide,

$$2\ Ph\text{-CH=O} \quad \xrightarrow[\text{H}_2\text{O}]{\text{KCN}} \quad Ph\text{-}\overset{\overset{\displaystyle OH}{|}}{CH}\text{-}\underset{\underset{\displaystyle O}{\|}}{C}\text{-}Ph$$

and the rate law is   rate $= k[\text{PhCHO}]^2\,[\text{CN}^-]$.

Provide a mechanism and demonstrate that it leads to the observed rate law. What features of cyanide as an anion and substituent make it such an effective catalyst for this reaction?

4.2 The Aldol reaction may be carried out under (i) specific base catalysis conditions, and (ii) general base catalysis conditions in deuterium oxide. In each case, where will deuterium be incorporated in (a) the product, and (b) unreacted ethanal isolated part way through the reactions?

4.3 The enolisation of simple ketones in aqueous solution buffered at $pH < 7$ by a weak acid AH and its sodium salt Na^+A^- is believed to proceed via rapid reversible protonation of the oxygen of the ketone by H_3O^+ and AH followed by rate-limiting deprotonation from the α-carbon by either water or A^-. Show that this mechanism leads to a general acid catalysis rate law with $k_o = 0$.

4.4 The base-catalysed decomposition of nitramide is important in the history of catalysis and one of the reactions originally investigated by Brønsted.

$$NH_2NO_2 \xrightarrow[H_2O]{B, OH^-} N_2O + H_2O$$

Provide a mechanism and show it leads to the general base catalysis rate law with $k_o = 0$.

4.5 The following results were obtained for the hydrolysis of phenyl vinyl ether catalysed by perchloric acid in aqueous solution at 25°C.

$$Ph\text{-}O\text{-}CH{=}CH_2 + H_2O \xrightarrow[H_2O]{H_3O^+} PhOH + CH_3CHO$$

10^2 [HClO$_4$]/mol dm^{-3}	$10^4 k_{obs}/s^{-1}$
2.96	0.96
3.94	1.28
5.91	1.97
7.97	2.60
10.2	3.41
10.4	3.37

Determine by a graphical method the catalytic constant for this reaction and propose a mechanism.

4.6 The hydrolysis of phenyl vinyl ether in aqueous solution at 25°C featured in problem 4.5 is also catalysed by general acids, AH.

Acid, AH	pK_a(AH)	$10^6 k_{AH}/dm^3$ mol^{-1} s^{-1}
ClCH$_2$CO$_2$H	2.87	40.4
PhOCH$_2$CO$_2$H	3.17	32.4
MeOCH$_2$CO$_2$H	3.57	10.2
HCO$_2$H	3.75	7.44
PhCH$_2$CO$_2$H	4.31	2.63
MeCO$_2$H	4.76	1.21

Is the mechanism provided in your answer to problem 4.5 also compatible with the results provided here? From these results and the catalytic constant determined for H_3O^+ in problem 4.5, determine the Brønsted α. What does the magnitude of α suggest regarding the mechanism?

4.7 The reaction of a methyl ketone with bromine in buffered aqueous solution is catalysed by a family of bases:

$$\underset{R}{\overset{O}{\underset{}{\|}}}\overset{}{C}{-}CH_3 \ + \ Br_2 \ \xrightarrow[\ -H^+\ ]{\ B/BH^+,\ H_2O\ } \ \underset{R}{\overset{O}{\underset{}{\|}}}\overset{}{C}{-}CH_2Br \ + \ Br^-$$

$$-d[\text{ketone}]/dt \ = \ k_{\text{obs}}[\text{ketone}] \qquad \text{where } k_{\text{obs}} \ = \ k_{OH}[OH^-] \ + \ k_B[B]$$

Provide a mechanism which accounts for the general base catalysis rate law and zero-order dependence upon bromine concentration. An analogous reaction occurs with iodine; on the basis of your mechanism, comment upon the probable rate law and relative magnitudes of the rate constants k_{OH} and k_B in the two halogenations.

4.8 Determine β_{nuc} for the hydrolysis of benzenesulfonyl chloride catalysed by substituted pyridines in aqueous solution at 25°C from the results below.

$$PhSO_2Cl \ + \ 2\,OH^- \ \xrightarrow[\ H_2O,\ k_{\text{obs}}\ ]{\ XC_5H_4N\ } \ PhSO_3^- \ + \ Cl^- \ + \ H_2O$$

$-d[PhSO_2Cl]/dt = k_{\text{obs}}[PhSO_2Cl]$ where $k_{\text{obs}} = k_o + k_{OH}[OH^-] + k_B[XC_5H_4N]$

Substituent, X	$k_B/dm^3\ mol^{-1}\ s^{-1}$	pK_a of $XC_5H_4NH^+$
3-Me	5.65	5.68
H	3.08	5.21
4-acetyl	0.54	3.48
4-CN	0.095	1.90
3-CN	0.053	1.39

Provide two mechanisms for the reactions catalysed by the pyridines. Which of your mechanisms is compatible with the result $k_B = 2.80 \ dm^3 \ mol^{-1} \ s^{-1}$ for the parent (X = H) in D_2O in place of H_2O.

4.9 Add paths and contours to the reaction map template on p. 74 to describe acyl transfers with
(a)   a concerted synchronous mechanism,
(b)   alternative asynchronous concerted mechanisms,
(c)   an associative step-wise addition-elimination mechanism, and
(d)   a dissociative step-wise mechanism via an acylium cation.

Discuss structural modifications in terms of perpendicular effects upon the transition structure which may be expected to convert a synchronous concerted mechanism into alternative asynchronous concerted then step-wise mechanisms.

By sketching Brønsted-type plots for the mechanisms (a), (c), and (d), discuss how these mechanisms could be distinguished.

5   Isotope effects upon organic chemical reactions

5.1 Introduction

In earlier chapters, we saw that mechanistic information may be gained by intensive study of a single particular reaction, e.g. by kinetics. We then saw that investigation of the effects of substituents in the substrate, or of changes to the structure or nature of a co-reactant such as a catalyst, leads to mechanistic information but about a family of reactions rather than a single one. Between these extremes of ever more detailed scrutiny of one reaction and investigating a widening family of related reactions lies the study of the effect of isotopic substitution upon reactants in a single reaction. On the one hand, replacement of one isotope by another, e.g. protium by deuterium, does not alter the chemical identity of the element and, consequently, the natures of the reactions of compounds remain unchanged. In other words, the potential energy hypersurface of the chemical system is independent of the pattern of isotopic substitution. On the other hand, isotopic substitution does alter the masses of groups within molecules and hence vibrational frequencies. As we saw in chapter one, chemical reactions have their origins in molecular vibrations; consequently, whilst isotopic substitution does not alter the nature of a chemical reaction, it may alter the rate at which the reaction occurs.

The second-order rate law of a generic bimolecular reaction,

$$A \; + \; B \; \rightarrow \; Product,$$

has the form
$$rate \; = \; k[A][B].$$

Clearly, if substitution of one isotope for another within either A or B alters what we above loosely called the rate of the reaction, it must do so by changing the value of the rate constant, k. This is true regardless of the order or nature of the reaction. In complex reactions under specified experimental conditions, isotopic substitution can change the magnitudes of elementary rate constants and, depending upon the mechanism of the overall transformation, thereby affect the overall experimental rate constant.

On the basis that an equilibrium is a balance between forward and reverse kinetic processes, we may also expect that isotopic substitution will alter the value of an equilibrium constant, but not the nature of the equilibrium. This indeed is observed. Before proceeding further, however, we need to define and explain a few terms.

Compounds which differ only in their isotopic composition, e.g. CH_4, CH_2D_2, and CD_4 are sometimes called isotopologues. Compounds of the same isotopic composition but with different isotopic distributions within the molecules, e.g. 1,1-dideuterioethene, and *cis*- and *trans*-1,2-dideuterioethene, are sometimes called isotopomers.

L is used for any isotope of hydrogen – protium, deuterium, or tritium. Hydron is the generic term for $^1H^+$, $^2H^+$, and $^3H^+$, i.e. L^+.

Definitions

Equilibrium isotope effect. The equilibrium constant for the hydron transfer reaction shown in eqn 5.1 has a particular value at 25°C in chlorobenzene as solvent when L = ^1H and a slightly smaller value when L = ^2H (or D).

$$K^H/K^D = 1.4 \quad (C_6H_5Cl, 25°C)$$

The modern convention is to express an isotope effect as the ratio of the value of the parameter (an equilibrium constant here) for the compound with the lighter isotope (protium in this case) over that for the compound with the heavier isotope (deuterium), i.e. K^H/K^D. When this ratio is greater than unity, i.e. when $K^H > K^D$, it is called a *normal* isotope effect.

Kinetic isotope effect. The second-order rate constant for the cycloaddition reaction of eqn 5.2 in toluene at 100°C is smaller when L = H than when L = D. Even though $k^H < k^D$, this kinetic isotope effect (i.e. a ratio of rate constants) is still written as k^H/k^D but now, because its value is less than unity, it is referred to as an *inverse* effect.

Usually, a deuterium isotope effect investigation involves first synthesis of the compound with natural abundance isotopic composition. The compound with a particular hydrogen completely replaced by a deuterium atom then has to be synthesised. The deuterium has to be introduced exactly and exclusively where it is required, and with as high a degree of incorporation as possible. Measurements are then normally made on the protiated and deuteriated samples at the same time and under identical experimental conditions by normal methods.

$$k^H/k^D = 0.95 \quad (PhCH_3, 100°C)$$

Primary isotope effect. When the chemical reaction involves cleavage of a bond to the isotopically labelled atom, the effect of the isotopic substitution is called a *primary* isotope effect. An equilibrium example is shown in eqn 5.1, and eqn 5.3 is a kinetic example in which hydroxide abstracts one of the weakly acidic methylene hydrogens of the ethyl nitroethanoate.

$$k^H/k^D = 4.6 \quad (water, 25°C)$$

Secondary isotope effect. When the bond to the isotopically labelled atom is not broken in the chemical reaction, we have a *secondary* isotope effect. The reaction of eqn 5.2 is a kinetic secondary isotope effect, and eqn 5.4 is an equilibrium example. Secondary isotope effects are almost invariably smaller than primary ones and, generally, the further the labelled atom is from the reaction site, the smaller the effect.

$$LCO_2H + H_2O \underset{H_2O}{\overset{K}{\rightleftharpoons}} LCO_2^- + H_3O^+ \qquad (5.4)$$

$$K^H/K^D = 1.08 \quad (\text{water, } 25°C)$$

Solvent deuterium isotope effect. Although, in principle, any atom in any solvent could be replaced by an isotope, and the effect of that substitution upon a reaction taking place in the solvent could be investigated, in practice, the effect is invariably the replacement of protium by deuterium. And because deuterium oxide is a relatively inexpensive chemical, the overwhelming majority of solvent isotope effect measurements have been in water, i.e. the effect of replacing H_2O by D_2O, as exemplified by eqn 5.5.

$$k^{H_2O}/k^{D_2O} = 1.20, \; 25°C.$$

Caveat! In spite of the implications of the preceding sections, it is not always a simple matter in practice to identify an experimental isotope effect as being of just one sort or another. Frequently, the overall effect of isotopic substitution is the result of various contributions as the following illustrates:

$$2\,L_2O \rightleftharpoons L_3O^+ + OL^-$$

$$K^H/K^D = 7.4 \qquad K^H/K^T = 16.4 \quad (25°C),$$

i.e., both ^2H_2O and ^3H_2O (T_2O) are much less dissociated than ^1H_2O. There will be primary, secondary, and solvent kinetic isotope effect contributions to forward and reverse reactions, the overall equilibrium effect being what is measured experimentally.

Similarly, weak acids are less dissociated in D_2O than in H_2O, and this is also a composite effect:

$$CH_3CO_2L + L_2O \rightleftharpoons CH_3CO_2^- + L_3O^+ \quad ; \; K_a$$

$$K_a^{H_2O}/K_a^{D_2O} = 3.33 \text{ at } 25°C.$$

5.2 Deuterium isotope effects upon molecular vibrations

X–L stretching vibrations

Figure 5.1 shows the anharmonic molecular potential energy curve for the stretching vibration of a diatomic molecule X–L where X is an atom much heavier than hydrogen.

Notice that the equilibrium isotope effects here are large (and hence must be principally primary effects), but they alone tell us nothing about the sizes of the individual equilibrium constants which, in these cases, are very small.

By considering the reverse, we see that D_3O^+ in D_2O is a stronger acid (hydron donor) than H_3O^+ in H_2O.

The energy difference between the protium and deuterium levels for a given vibrational quantum number decreases as the quantum number increases and the curve becomes increasingly anharmonic. It becomes zero upon dissociation, i.e. when the vibration has become a translation.

Note that although XH and XD molecules share the same potential energy curve, the time-averaged bond lengths of the two molecules are slightly different. That of XH is slightly longer than that of XD; this is a vibrational consequence of the asymmetry of the curve due to anharmonicity.

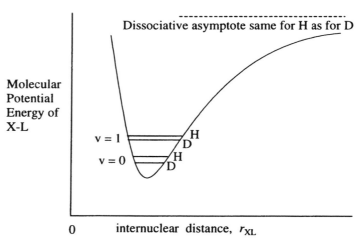

Fig. 5.1 Unsymmetrical anharmonic potential energy curve for the stretching vibration of molecule X–L showing X–D levels lower than X–H levels for v = 0 and v = 1

For very small displacements from the equilibrium bond length, i.e. close to the minimum, the curve is approximately harmonic and the vibrational frequency, v, is given by eqn 5.6

$$v = \frac{1}{2\pi}\sqrt{\frac{\kappa}{\mu_{XL}}} \tag{5.6}$$

The force constant of the bond is determined by the electronic configuration of the molecule, i.e. by the elements involved (hydrogen and X) but it is not affected by the isotopes of the elements.

The reduced mass is given by
$$\frac{1}{\mu_{XL}} = \frac{1}{m_X} + \frac{1}{m_L}$$
or $$\mu_{XL} = \frac{m_X \cdot m_L}{(m_L + m_X)}$$
where m_i = mass of i.

where κ = the force constant for the X–L bond stretching vibration, and
μ_{XL} = the reduced mass of the X–L molecule.

Since the reduced mass is different for protium and deuterium analogues, they will have different vibrational frequencies and the quantised vibrational energies of the two molecules are given by

$$E = (v + 0.5)hv$$

where v is the vibrational quantum number included in Fig. 5.1.

The energy, therefore, depends upon the frequency of the vibration which, by eqn 5.6, depends upon the reduced mass of the molecule, i.e. the masses of the two atoms. Consequently, the energies of protium and deuterium analogues in the same vibrational state (same v) are different, that for deuterium being lower than that for protium. This is illustrated for the zero and first quantised vibrational levels (v = 0 and v = 1) in Fig. 5.1.

Y–L bending vibrations

The number and nature of the bending vibrations of the Y–L group depend upon the symmetry and nature of the molecule as a whole.

A diatomic molecule X–L, for example when X = Br or I, has only a single internal degree of freedom, the stretching vibration considered above. However, if a hydrogen is bonded to a multivalent atom, e.g. carbon, nitrogen, or oxygen, which is part of a larger group, Y, there can also be Y–L bending vibrations in addition to the stretching. The limit of such a vibration as the amplitude increases, however, does not lead to a dissociation, and a potential energy curve describing a bending vibration is illustrated in Fig. 5.2. Again, quantised levels are included with deuterium ones lower than protium ones.

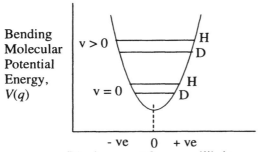

Fig. 5.2 Potential energy curve for a bending vibration of molecule Y–L showing D and H levels for v = 0 and v > 0

As for a stretching vibration, a bending vibration has a force constant and the bending displacement of L from its equilibrium position is described by a unidimensional structural variable, q. Furthermore, the mathematical relationship between the potential energy of the molecule and q is still approximately simple harmonic at small displacements, and anharmonicity increases as the amplitude of the vibration increases. However, Fig. 5.2 is more symmetrical than Fig. 5.1 for a stretching vibration. In the former, the effects of anharmonicity are similar for positive and negative displacements whereas, in the latter, the effect of anharmonicity for stretching the bond is opposite to the effect for compression.

> Bending force constants are much smaller than stretching ones.

5.3 Primary deuterium kinetic isotope effects

The origin of primary deuterium kinetic isotope effects

An over-simplistic model for hydrogen transfer between two groups, eqn 5.7, involves A–H bond cleavage, then rebonding of the free hydrogen to B.

$$\text{A-H} + \text{B} \rightleftharpoons \text{A} + \text{H-B} \qquad (5.7)$$

> The hydrogen could be transferred as a hydron, hydrogen atom, or hydride. Consequently, here and below we do not include charge.

A more realistic model (now replacing H by L, eqn 5.8) involves molecules AL and B coming together to form an encounter complex of the appropriate configuration; the hydrogen transfer then takes place to give an isomeric complex, and the new molecules A and LB separate.

$$\text{A-L} + \text{B} \rightleftharpoons \text{A-L}\,.\,\text{B} \rightleftharpoons \text{A}\,.\,\text{L-B} \rightleftharpoons \text{A} + \text{L-B} \qquad (5.8)$$

> AL and B could come together in any number of ways, but only the collinear arrangement A-L····B is included in Fig. 5.3.

To plot the energetics of this overall process, we need an energy coordinate and two types of configurational coordinates. The intermolecular distances initially between AL and B, and later between A and LB, are translational degrees of freedom. The position of the hydrogen between A and B in the encounter complex, however, is within a vibrational degree of freedom.

Because the intermolecular motion between molecules is much slower than vibrational motion within a molecule, we may separate the two. First, AL and B come together, then the hydrogen transfer from A to B takes place within the encounter complex with A and B at virtually a fixed distance apart. Just this second aspect, i.e. the transfer of L from A to B at a fixed A - B internuclear distance is shown in Fig. 5.3. As we saw in the first chapter, this comprises the intersecting anharmonic potential energy profiles for the A–L and L–B stretching vibrations.

This reaction profile does not include the coming together of AL and B, or the separation of A and LB.

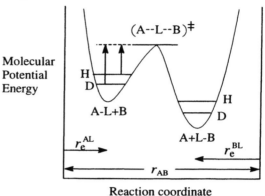

Fig. 5.3 Molecular potential energy profile for the hydrogen transfer from A to B within the encounter complex of the mechanism of eqn 5.8

A more detailed analysis of the origin of primary deuterium kinetic isotope effects includes the occupancies of the higher vibrational states as well as the energy difference between the zero point levels. It turns out, however, that the overall effect is dominated by the latter.

The point at which the two potential energy curves intersect corresponds to a configuration with the hydrogen bonded partly and equally to both A and B. This is the transition structure for the hydrogen transfer and the reaction coordinate at this point is again a translational degree of freedom rather than a vibration. Consequently, the structures with different hydrogen isotopes have the same energy as indicated by the single horizontal line. We see, therefore, that the kinetic isotope effect (k^H/k^D) is determined by the different barrier heights from the H and D levels in A–L, i.e. the difference in the zero-point energies of AH and AD, $\{\varepsilon_0(AH)- \varepsilon_0(AD)\} = \delta\varepsilon_0$.

Primary deuterium kinetic isotope effects and reaction mechanisms

According to quantum theory, the molecular zero-point energy of a stretching vibration is given by eqn 5.9 where ν is the vibrational frequency.

$$\varepsilon_0 = 0.5h\nu \qquad (5.9)$$

More advanced theory of hydrogen transfer invoking quantum mechanical tunnelling is required to account for some very much larger primary deuterium kinetic isotope effects which have been reported.

For hydrogens bonded to carbon, the stretching vibrational frequency is easily calculated from the wavenumber of the C–L stretch in the infra-red spectrum ($2914cm^{-1}$ for C–H and $2085cm^{-1}$ for C–D). From these results, we calculate that the difference in zero-point energies for C–H and C–D, $\delta\varepsilon_0$, scaled up to the molar level, is 4.96 kJ mol^{-1}, and $k^H/k^D \sim 7.5$ at 25°C. We conclude from this simple analysis that a chemical reaction which proceeds via rate-limiting cleavage of a carbon-hydrogen bond will have an appreciable primary deuterium kinetic isotope effect. Results for reactions which, by other evidence, are believed to react by such mechanisms range between $k^H/k^D \sim 3$ and 10 at 25°C, and the following E2 reaction is representative.

EtO⁻ + L—C(Ph)(CH₃)—CH₂Br $\xrightarrow[\text{25°C}]{\text{EtOH}}$ (Ph)(CH₃)C=CH₂ + EtOL + Br⁻

$k^H/k^D = 7.5$

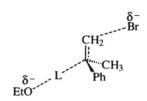

Appreciably smaller primary deuterium kinetic isotope effects are observed in E2 reactions if the transition structure is very unsymmetrical with respect to hydron transfer.

The very substantial deuterium kinetic isotope effects in the reactions of eqn 5.10 corroborate other evidence (general base catalysis rate law and zero-order dependence on the bromine concentration) that they proceed via rate-limiting hydron abstraction from the ketones to give enolates which then react very rapidly with the halogen to give α-bromoketones.

$$R-CO-CL_3 + Br_2 + OH^- \xrightarrow[\text{25°C}]{H_2O} R-CO-CL_2-Br + Br^- + HOL \quad (5.10)$$

k^H/k^D ca. 7

One of the earliest uses of deuterium kinetic isotope effect measurements to investigate mechanism was in the oxidation of 2-propanol with acidified potassium dichromate (chromic acid), Fig. 5.4. The magnitude of the isotope effect is too large to be other than primary, i.e. the α-C–L bond is undergoing cleavage in the rate-limiting step of a complex mechanism involving initial formation of an intermediate chromate ester.

(CH₃)(CH₃)C(L)OH $\xrightarrow[\text{K}_2\text{Cr}_2\text{O}_7,\ 40°C]{H_2O,\ H_3O^+}$ CH₃—CO—CH₃

$k^H/k^D = 5.9$

:OH₂

(CH₃)(CH₃)C(L)—O—CrO₃H

This oxidation reaction may be regarded as an elimination related to the E2 mechanism by which alkenes are formed, but here the double bond is formed between carbon and oxygen.

Fig. 5.4 Primary deuterium kinetic isotope effect in the oxidation of 2-propanol by chromic acid in aqueous solution

5.4 Secondary deuterium kinetic isotope effects

The origin of secondary deuterium kinetic isotope effects

Occasionally, isotopic substitution away from the reaction site leads to a small rate effect. Whilst no rebonding to the labelled atom occurs in the transition state of such a reaction, there must be a change in molecular vibrations involving the isotopically substituted atom as reactant molecule becomes

A ⇌ A‡ ⟶ Product

At the maximum in the reaction coordinate, the transition structure has real vibrations in all normal modes except the reaction coordinate. In that particular degree of freedom, there is an imaginary frequency corresponding to a negative force constant.

transition structure. This is illustrated in Fig. 5.5 for a unimolecular reaction of molecule A via transition structure $A^{\ddagger}$.

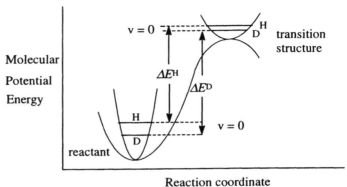

Fig. 5.5 Origin of a secondary deuterium kinetic isotope effect in a unimolecular reaction

In this process, the reaction coordinate is initially a normal molecular vibration of the reactant represented by the larger minimum to the left in Fig. 5.5. There is then a maximum in the reaction coordinate at $A^{\ddagger}$; consequently, the reaction coordinate cannot be a normal molecular vibration at this point.

Superimposed upon the reaction coordinate are two other potential energy curves which should be seen as in and out of the plane of the paper; one is for the reactant, the other for the transition structure. The former represents a molecular vibration involving the isotopically labelled atom (stretching or bending), and the separate deuterium and protium zero-point energy levels are included. The latter represents the corresponding vibration in the transition structure, and again protium and deuterium zero-point energy levels are included. Thus, as reactant molecule becomes transition structure, a vibration, which is not a component of the reaction coordinate but which involves the labelled atom, undergoes a transformation. The extent to which this vibration is different between reactant and transition structure will depend upon the nature of the vibration, its remoteness from the reaction site, and the nature of the actual reaction. In Fig. 5.5, the potential energy curve for this vibration is narrower (larger force constant and more widely spaced vibrational levels) in the reactant than in the transition structure, but it could be otherwise.

If a vibration does not change as reactant molecule becomes transition structure, e.g. because it is very remote from the reaction site, then differences between protium and deuterium levels will be the same in reactant and transition structure, and this vibration will not contribute to a secondary deuterium kinetic isotope effect.

The rate constant at a particular temperature depends upon the potential energy difference between reactant and activated complex (amongst other factors), and this is usually represented as the barrier in the potential energy reaction profile. More precisely, it is the zero-point energy difference between reactant molecule and transition structure summed over all internal degrees of freedom, i.e. including the ones involving the isotopic atom. Consequently, if the energy differences between zero-point deuterium levels of reactant and those of transition structure (ΔE^{D} in Fig. 5.5 for just one such vibration) are not the same as the corresponding differences between protium levels (ΔE^{H} in

Fig. 5.5 for just one), the activation energies will not be the same and the rate constants for protium and deuterium analogues will be different. Hence, we can have isotope effects upon unimolecular chemical reactions even when the bonds to the isotopically labelled atoms remain intact, but the effects are much smaller than primary isotope effects.

Secondary deuterium kinetic isotope effects and reaction mechanisms

Unimolecular mechanisms. For a reaction described by Fig. 5.5 in which there is a reduction in the force constant of the vibration as reactant becomes transition structure, there will be a normal kinetic isotope effect, i.e. $k^H/k^D > 1$. This is observed when a carbon bearing the isotopic hydrogen changes from sp^3 to sp^2 as for the α-carbon in the rate-limiting step of an $S_N1/E1$ mechanism, e.g. eqn 5.11. Indeed, in substitution/elimination reactions with oxygen-bonded nucleofuges, an α-deuterium kinetic isotope effect of ca. 1.2 is reliable evidence of an $S_N1/E1$ mechanism.

$$\text{Me}_{\text{...}}\underset{\text{Me}}{\overset{\text{L}}{\diagdown\diagup}}\text{OTs} \quad \xrightarrow[\text{25}^\circ\text{C}]{\text{CF}_3\text{CO}_2\text{H}} \quad \text{Solvolysis, } S_N1/E1 \qquad (5.11)$$
$$k^H/k^D = 1.22$$

On the other hand, if the parabola describing the C–L vibration in the transition structure is narrower than that for its vibration in the reactant, i.e. there is an increase in the force constant of the C–L vibration as reactant becomes transition structure, then an inverse deuterium kinetic isotope effect will be observed, i.e. $k^H/k^D < 1$. This is typical of reactions in which the carbon bearing the hydrogen(s) undergoes a change from sp^2 to sp^3 and an example is shown in eqn 5.12 although the effect here is rather small.

$$\begin{array}{c}\text{O} \diagup \text{CH}_2 \\ \text{H}_2\text{C} \diagdown \diagup \text{CL}_2\end{array} \quad \xrightarrow[k^H/k^D = 0.976]{\text{160}^\circ\text{C}} \quad \begin{array}{c}\text{O} \diagup \text{CH}_2 \\ \text{H}_2\text{C} \diagdown \diagup \text{CL}_2\end{array} \qquad (5.12)$$

Bimolecular mechanisms. The analysis of secondary deuterium kinetic isotope effects based upon Fig. 5.5 may be extended to include bimolecular mechanisms and the deductions are similar. If the force constant of a molecular vibration involving a C–L bond (which is not part of the reaction coordinate) undergoes a change as reactants become transition structure, then a finite deuterium kinetic isotope effect is expected. If the carbon bearing the hydrogen undergoes an increase in coordination as reactant becomes transition structure, effects are usually inverse as in the following addition reaction.

$$\underset{\text{Ph}}{\overset{\text{O}}{\overset{\|}{\diagup\diagdown}}}\text{L} \quad + \text{ CN}^- \quad \xrightarrow[\text{25}^\circ\text{C}]{\text{H}_2\text{O}} \quad \underset{\text{Ph}_{\text{...}}}{\overset{\text{O}^-}{\diagup\diagdown}}\underset{\text{L}}{\overset{}{\diagdown}}\text{CN} \qquad k^H/k^D = 0.833$$

More often, however, effects are small as in the example of eqn 5.2 (p. 82).

In S_N2 reactions when the isotopic hydrogen is on the α-carbon, effects are also small with oxygen bonded nucleofuges and nucleophiles but often normal ($k^H/k^D > 1$).

$$\underset{CL_2}{\overset{Me}{\diagdown}}\!\!\!\diagup OTs \;+\; 2\,H_2O \quad \xrightarrow[\,54.3^oC\,]{\,H_2O\,} \quad \underset{CL_2}{\overset{Me}{\diagdown}}\!\!\!\diagup OH \;+\; H_3O^+ \; TsO^-$$

$$k^H/k^D = 1.038 \text{ for the two D atoms}$$

If there is a decrease in coordination at the carbon bearing the isotopic hydrogen, as in an α,β-elimination reaction of a substrate with an α-C–L, then a normal effect is anticipated, i.e. $k^H/k^D > 1$, as in the following E2 reaction.

$$k^H/k^D = 1.14$$

Again, however, effects are seldom more than 20% and often less than 5%.

5.6 Heavy atom kinetic isotope effects

Isotope effects for elements other than hydrogen have been determined in mechanistic investigations but, as the ratio of rate constants depends upon the ratio of isotopic masses (rather than their difference), much smaller effects are observed. Because the effects are so small and it is difficult to obtain isotopically pure samples, normal kinetics procedures are inadequate. Rate constants k^H and k^D are determined by separate (but usually concurrent) rate measurements upon the ^1H and ^2H analogues, but heavy atom isotope effects are measured by a quite different procedure. A sample of the reactant with an accurately known ratio of isotopes at a particular location, e.g. ^{12}C and ^{13}C or ^{14}N and ^{15}N, is reacted then the ratio is again accurately measured mass spectrometrically. From the ratio before and after a precisely known extent of reaction, the ratio of rate constants can be calculated.

Benzyl halides undergo substitution reactions in aqueous solution, and two benzylic α-carbon kinetic isotope effects are shown below.

$$\underset{CH_2}{\overset{Ph}{\diagdown}}\!\!\!\diagup Cl \;+\; Y^- \quad \xrightarrow[\,60^oC\,]{\,H_2O\text{ - acetone}\,} \quad \underset{CH_2}{\overset{Ph}{\diagdown}}\!\!\!\diagup Y \;+\; Cl^-$$

$$k^{12}/k^{14} = 1.130 \text{ for } Y^- = N_3^- \quad k^{12}/k^{14} = 1.085 \text{ for } HY = H_2O$$

These results are in accord with the S_N2 mechanism established for such reactions; consequently, there should be heavy atom kinetic isotope effects in both nucleofuge and nucleophile. These have been observed in the following.

$$CH_3\text{–}I \;+\; CN^- \quad \xrightarrow[\,11.4^oC\,]{\,H_2O\,} \quad CH_3\text{–}CN \;+\; I^-$$

$$k^{12}/k^{13} = 1.0149 \text{ in } CN^-$$

The theory of heavy atom kinetic isotope effects is less well developed than for deuterium. Even so, merely the existence of an effect is sufficient to distinguish between some mechanisms.

Meaningful conclusions may be drawn from heavy atom kinetic isotope effects which are close to unity. Consequently, high accuracy is essential but not easy to achieve.

Experimental errors are not given in results quoted here but they are given in the original literature and are essential in assessing whether a measured effect is real.
For carbon, ^{12}C may be substituted by ^{13}C or ^{14}C. The latter leads to larger isotope effects due to its larger mass, but is radio-active so requires special facilities.

$$\text{Ar-CH}_2\text{-Cl} \ + \ \text{CN}^- \ \xrightarrow[\text{30°C}]{\text{H}_2\text{O - dioxane}} \ \text{Ar-CH}_2\text{-CN} \ + \ \text{Cl}^-$$

$$k^{35}/k^{37} = 1.0057 \text{ at } \text{Cl for Ar} = 4\text{-NO}_2\text{C}_6\text{H}_4$$

In S_N1 substitution mechanisms, even smaller (but real) kinetic isotope effects are observed at carbon, but larger ones in nucleofuges.

5.7 Solvent deuterium kinetic isotope effects

The title implies a generality which is seldom observed; overwhelmingly, the effect is of changing the solvent from H_2O to D_2O.

Distinguishing between nucleophile catalysis and mechanistic general base catalysis

Since hydron transfer is involved in mechanistic general base catalysis, this mechanism will lead to an appreciable effect ($k^{H_2O}/k^{D_2O} >$ ca. 2) as in the ethanoate-catalysed hydrolysis of ethyl difluoroethanoate, eqn 5.13.

$$\text{CHF}_2\text{CO}_2\text{Et} \ + \ \text{L}_2\text{O} \ \xrightarrow[\text{25°C}]{\text{AcO}^-} \ \text{CHF}_2\text{CO}_2\text{L} \ + \ \text{EtOL} \qquad (5.13)$$

$$k_{\text{AcO}}(\text{H}_2\text{O})/k_{\text{AcO}}(\text{D}_2\text{O}) = 2.9$$

In contrast, no solvent hydron is transferred in the rate-limiting step of nucleophile catalysis so, if this is the mechanism, there will be a smaller effect ($k^{H_2O}/k^{D_2O} <$ ca. 2) as in the reaction of eqn 4.21 on p. 68.

Distinguishing between pre-equilibrium and rate-limiting protonation

Catalysis of a reaction by hydronium ion could be due to either pre-equilibrium protonation of the substrate and a subsequent rate-limiting step, or initial rate-limiting protonation of the substrate. If the reaction is specific acid catalysed (catalysis by H_3O^+ but not by general acids), it cannot involve rate-limiting proton transfer. If the reaction is general acid catalysed, however, there is still an ambiguity. It could involve initial rate-limiting protonation of the substrate, or pre-equilibrium protonation followed by rate-limiting *deprotonation* from elsewhere in the molecule (as in the acid-catalysed enolisation of simple ketones). The solvent deuterium kinetic isotope effect for hydronium ion catalysis distinguishes between

(i) *pre-equilibrium* hydron transfer to the substrate and a subsequent rate-limiting step (regardless of the nature of the rate-limiting step), and

(ii) *rate-limiting* initial hydron transfer to substrate.

Figure 5.6 is an outline of one type of the former, i.e. pre-equilibrium hydron transfer and a rate-limiting step *not* involving hydron transfer (so this reaction will have a specific acid catalysis rate law). K is the equilibrium constant for the pre-equilibrium and k is the elementary rate constant for the rate-limiting step (which could include another reactant).

The higher mass of deuterium and the vibrational consequences lead to differences in physical properties of H_2O and D_2O. For example, D_2O is higher boiling and more viscous than H_2O. The higher viscosity causes second-order rate constants in D_2O, including the diffusion controlled limit, to be lower than in H_2O.

This topic is also discussed in chapter 4.

A specific acid catalysis rate law requires pre-equilibrium protonation followed by a rate-limiting step *not* involving proton transfer, see p. 54.

$$S + L_3O^+ \rightleftharpoons SL^+ + L_2O \; ; K$$

$$SL^+ + L_2O \xrightarrow{k} Product + L_3O^+$$

Fig. 5.6 Pre-equilibrium hydron transfer followed by a rate-limiting step *not* involving hydron transfer – specific acid catalysis and an inverse H_2O/D_2O kinetic isotope effect

The rate of this reaction is limited by the rate of the second step, i.e. by the magnitude of k and the concentration of SL^+. In turn, $[SL^+]$ is determined by the magnitude of K for the first step, and the concentrations of S and L_3O^+. The concentration of L_3O^+ and the initial concentration of S can be identical in the reactions in H_2O and D_2O, so any rate effect upon changing from H_2O to D_2O must be caused by changes in K and k.

We saw on p. 83, that the dissociation constant of a weak acid is smaller in D_2O than in H_2O, and the first step in Fig. 5.6 is the reverse of the dissociation of a weak acid, SL^+. Consequently, K for the reaction shown will be *larger* in D_2O than in H_2O. The rate constant for the second step, k, which does not involve a hydron transfer, should be much the same in the two solvents. The mechanism of Fig. 5.6, therefore, which leads to a specific acid catalysis rate law, will also lead to an *inverse* solvent deuterium kinetic isotope effect ($k^H/k^D < 1$ where k^L is the experimental catalytic constant for catalysis by L_3O^+ in L_2O). This has been very widely observed and the hydrolysis of simple acetals, eqn 5.14 (R, R' = alkyl), is representative.

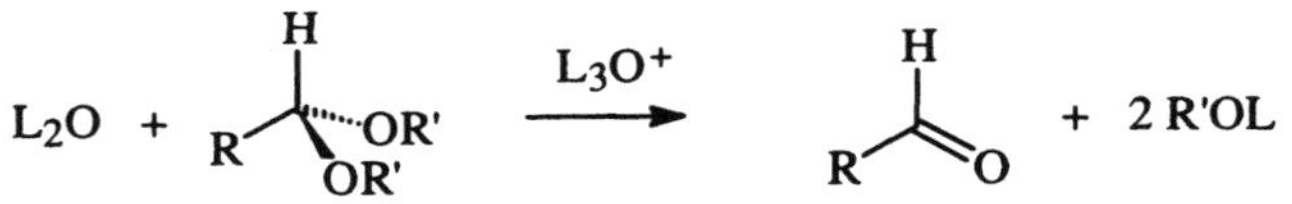

D₃O⁺ is a better hydron donor (stronger acid) than H₃O⁺.

The solution of D₃O⁺ is normally generated by dilution in D₂O of commercially available concentrated perchloric acid in D₂O, i.e. D₃O⁺ ClO₄⁻.

The general acid catalysed enolisation of simple ketones involves pre-equilibrium protonation on oxygen then rate-limiting proton abstraction from the α-carbon. This mechanism also leads to an inverse solvent deuterium kinetic isotope effect for catalysis by hydronium ion (*k*ᴴ/*k*ᴰ ca. 0.5 for acetone).

$$L_2O + \underset{R}{\overset{H}{\underset{\displaystyle OR'}{\overset{\displaystyle \cdots OR'}{\diagup\!\!\!\diagdown}}}} \xrightarrow{L_3O^+} \underset{R}{\overset{H}{\diagdown}}C\!=\!O + 2\,R'OL \qquad (5.14)$$

$$k^H/k^D = 0.3 - 0.5 \; (25°C)$$

In contrast, the general acid catalysis mechanism involving initial rate-limiting hydron transfer will be appreciably slower for D transfer than for H transfer for the reasons discussed earlier in this chapter. Consequently, the catalytic constant k^L for this mechanism will be *smaller* for D_3O^+ than for H_3O^+ (a normal kinetic isotope effect, $k^H/k^D > 1$) as shown in eqn 5.15.

The relatively modest normal solvent deuterium kinetic isotope effect for catalysis by L₃O⁺ in eqn 5.15 is due to partial compensation of a large normal primary effect by a substantial inverse secondary effect. The kinetic isotope effect for catalysis by HCO₂L in L₂O for which there is no compensating inverse secondary effect is *k*ᴬᴴ/*k*ᴬᴰ = 6.8.

$$CH_2\!=\!CH\text{-}OEt + L_2O \xrightarrow[25°C]{L_3O^+} LCH_2CH\!=\!O + EtOL \qquad (5.15)$$

$$k^H/k^D = 2.95$$

Enzyme catalysed reactions

Enzyme mechanisms have also been investigated by measurement of substrate isotope effects, especially secondary α-deuterium kinetic isotope effects in synthetic substrates for hydrolytic enzymes, and by heavy atom isotope effects.

Kinetic parameters obtained from an enzymic rate study seldom relate directly to individual steps of what is generally a complex overall reaction. Consequently, it is usually necessary to carry out experiments designed to allow the isotope effects for separate stages of the overall process to be determined. This is a specialised and difficult area, and we shall give only a single example which may not be representative.

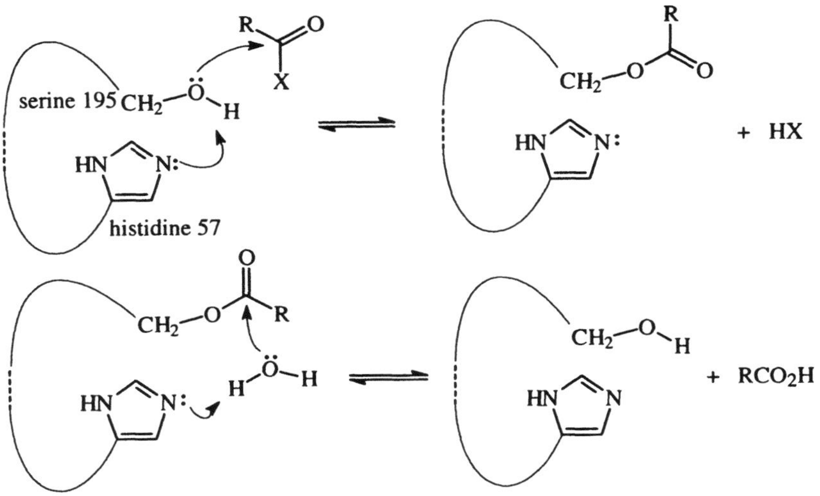

Fig. 5.7 Overall acylation and deacylation steps by mechanistic general base catalysis in the hydrolysis of amides and esters by α-chymotrypsin

The tetrahedral intermediates in both steps are not shown.

α-Chymotrypsin is an intestinal enzyme which hydrolyses specific peptide links in proteins, but it also catalyses the hydrolysis of non-natural substrates including esters and amides simpler than proteins. The mechanism involves acyl transfer from substrate to the hydroxyl of a serine unit of the enzyme, then its deacylation. The imidazole residue of a histidine unit of the enzyme is involved in both steps and, in principle, could act by mechanistic general base catalysis or nucleophile catalysis.

For a range of substrates, k^{H_2O}/k^{D_2O} results are between about 2 and 3 for both the acylation of the serine hydroxyl and the subsequent deacylation which implicates mechanistic general base catalysis in both steps of the acyl transfer. The overall process is outlined in Fig. 5.7 where X in the substrate is a nucleofuge bonded through either nitrogen or oxygen.

Problems

5.1 The α-bromination of simple methyl ketones with bromine in dilute aqueous strong acids is second order,

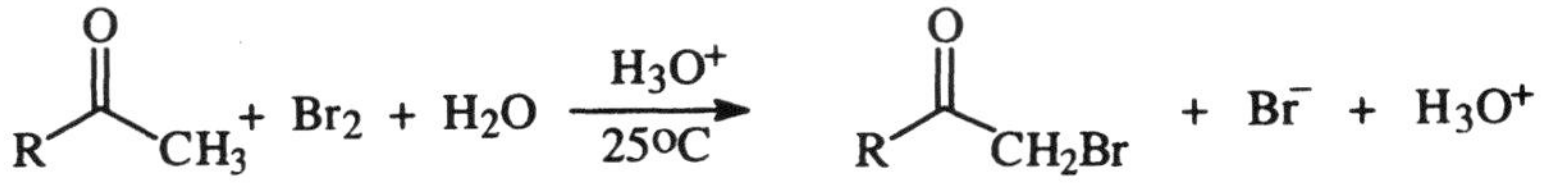

$$\text{rate} = k\,[\text{ketone}][\text{H}_3\text{O}^+],$$

and the kinetic effect of replacing CH_3 by CD_3 is

$$k^H/k^D = \text{ca. } 7.$$

(a) Provide a mechanism which accommodates these results.

(b) According to your proposed mechanism,
  (i) what would be the kinetic effect of replacing Br_2 by I_2?
  (ii) what would be the approximate ratio of catalytic rate constants in H_2O and D_2O?
(c) In the presence of a weak acid, will the reaction be specific or general acid catalysed?

5.2 The uncatalysed hydrolysis of ethanoic anhydride has an appreciable deuterium solvent kinetic isotope effect.

$$Ac_2O \; + \; L_2O \; \xrightarrow{\text{L}_2\text{O, 25°C}} \; 2 \; AcOL \; ; \quad k^{H_2O}/k^{D_2O} \; = \; 2.89$$

Formulate a mechanism compatible with this result.

5.3 In two aromatic electrophilic substitution reactions, the following rate and kinetic isotope effect results were obtained.

$$\text{rate} = k[\text{HOBr}][C_6L_6][H_3O^+] \; , \; k(C_6H_6)/k(C_6D_6) \; = \; 1.0$$

$$k^H/k^D \; = \; 1.4 \; (25°C)$$

(a) Give a mechanism for the first reaction compatible with the rate law.
(b) What information do the isotope effects provide about the mechanisms of the two reactions?

5.4 The hydrolysis of 2-(*p*-nitrophenoxy)tetrahydropyran (PNT) shown below occurs by uncatalysed and hydronium ion catalysed routes.

$$\text{rate} = (k_o \; + \; k_H[H_3O^+]) \; [\text{PNT}];$$

(a) Suggest possible mechanisms.
(b) What deductions can be made regarding the transition structures in the uncatalysed and catalysed reactions from the following α-deuterium secondary kinetic isotope effect measurements?

$$k_o^H/k_o^D = 1.17 \; (46°C); \quad k_H^H/k_H^D = 1.07 \; (20°C)$$

(c) Under the conditions of the reaction, the 2-tetrahydropyranol product is in equilibrium with an isomer. Propose a structure for this isomer and possible mechanisms for the isomerisation.

References and background reading

R. P. Bell, *The proton in chemistry* (2nd edn), Chapman and Hall, London (1973).

E. F. Caldin and V. Gold (editors), *Proton transfer reactions,* Chapman and Hall, London (1975).

B. K. Carpenter, *Determination of organic reaction mechanisms,* Wiley-Interscience, New York (1984).

N. B. Chapman and J. Shorter (editors) *Correlation analysis in chemistry: recent advances,* Plenum Press, New York (1978).

R. D. Guthrie and W. P. Jencks, "IUPAC recommendations for the representation of reaction mechanisms", *Acc. Chem. Res.,* **22**, 343 (1989).

L. P. Hammett, *Physical organic chemistry* (2nd edn), McGraw-Hill, New York (1970).

J. Hine, *Structural effects on equilibria in organic chemistry,* Wiley-Interscience, New York (1975).

C. K. Ingold, *Structure and mechanism in organic chemistry* (2nd edn), G. Bell and Sons, London (1969).

N. S. Isaacs, *Physical organic chemistry* (2nd edn), Longman, Harlow (1995).

W. P. Jencks, *Catalysis in chemistry and enzymology,* McGraw-Hill, New York (1969).

C. D. Johnson, *The Hammett equation,* Cambridge University Press, London (1973).

T. H. Lowry and K. S. Richardson, *Mechanism and theory in organic chemistry* (3rd edn), Harper Collins, New York (1987).

H. Maskill, *The physical basis of organic chemistry,* Oxford University Press, Oxford (1985).

H. Maskill, *Mechanisms of organic reactions,* Oxford University Press, Oxford (1996).

L. Melander and W. H. Saunders, *Reaction rates of isotopic molecules,* Wiley, New York (1980).

J. W. Moore and R. G. Pearson, *Kinetics and mechanism* (3rd edn), Wiley, New York (1981).

P. Müller, "Glossary of terms used in physical organic chemistry", *Pure & Appl. Chem.,* **66,** 1077 (1994).

M. I. Page and A. Williams (editors), *Enzyme mechanisms,* Royal Society of Chemistry, London (1987).

M. I Page and A. Williams, *Organic and bio-organic mechanisms,* Addison Wesley Longman, Harlow (1997).

C. Reichardt, *Solvents and solvent effects in organic chemistry* (2nd edn), VCH Weinheim, Germany (1988).

C. Walsh, *Enzymatic reaction mechanisms,* Freeman, San Francisco (1979).

In addition to the above, two very important series are
Advances in physical organic chemistry, Academic Press, London, and
Progress in physical organic chemistry, Wiley-Interscience, New York.

Index

Appendix

Useful data

N_A	Avogadro constant	6.022×10^{23} mol^{-1}
h	Planck constant	2.9979×10^{-34} J s
k_B	Boltzmann constant	1.381×10^{-23} J K^{-1}
R	gas constant	8.314 J K^{-1} mol^{-1}
e	exponential number	2.7183

Equivalencies

$0°C = 273.15$ K $\qquad\qquad$ 1 cal $= 4.184$ J

$\log_{10} z = \log_{10} e . \ln_e z = 0.4343 \ln_e z$ $\qquad$ $\ln_e z = \ln_e 10 . \log_{10} z = 2.303 \log_{10} z$

Important equations

$$k_c = \frac{k_B T}{h} . e^{-\Delta G^\ddagger / RT}$$

Arrhenius equation $\qquad$ $k = A . e^{-E_a/RT}$ $\quad$ and $\quad$ $\ln k_c = \ln A - \frac{E_a}{RT}$

$E_a = \Delta H^\ddagger + RT$ $\qquad$ $A = \frac{k_B T}{h} . e . e^{\Delta S^\ddagger / R}$

Eyring equation $\qquad$ $\ln\left(\frac{k_c}{T}\right) = \ln\left(\frac{k_B}{h}\right) + \frac{\Delta S^\ddagger}{R} - \frac{\Delta H^\ddagger}{RT}$

$\sigma_X = pK_a(C_6H_5CO_2H) - pK_a(XC_6H_4CO_2H)$

Hammett equation for equilibria $\qquad$ $\log\left\{\dfrac{K^X}{K^H}\right\} = \rho.\sigma_X$, $\;$ and for kinetics $\;$ $\log\left\{\dfrac{k^X}{k^H}\right\} = \rho.\sigma_X$

Yukawa-Tsuno equation for kinetics $\quad$ $\log\left\{\dfrac{k^X}{k^H}\right\} = \rho\{\sigma_X + r^+(\sigma_X^+ - \sigma_X)\}$

(also for equilibrium constants K, and with r^- and σ^- in place of r^+ and σ^+)

Brønsted equation $\quad$ for general acid catalysis $\qquad$ $\delta(\log k_{AH}) = -\alpha\, \delta(pK_a)$

$\qquad\qquad\qquad\quad$ for general base catalysis $\qquad$ $\delta(\log k_B) = \beta\, \delta(pK_a)$

$\qquad\qquad\qquad\quad$ for substitution reactions $\qquad$ $\delta(\log k_{nuc}) = \beta_{nuc}\, \delta(pK_a)$

$\qquad\qquad\qquad\qquad\qquad\qquad$ and $\qquad$ $\delta(\log k_{lg}) = -\beta_{lg}\, \delta(pK_a)$

Printed in the United States of America/BNB